Virtual Water

Virtual Water explores the role of "virtual water" – the water embedded in a product – in ongoing conversations of agriculture, trade and sustainability in an increasingly interconnected world.

A pervasive theme throughout the book is the general lack of knowledge of the use of water in producing and consuming food. The chapters, arising from a workshop supported by the OECD Co-operative Research Programme: Biological Resources Management for Sustainable Agricultural Systems, on virtual water, agriculture and trade at the University of Nebraska-Lincoln, consider questions of gaps in knowledge, why sustainability matters and the policy implications of virtual water trade. Contributors show how water is a lens through which to examine an array of vital issues facing humanity and the planet: human and animal health; food production; environmental management; resource consumption; climate change adaptation and mitigation; economic development, trade and competitiveness; and ethics and consumer trust.

Virtual Water will be of great interest to scholars of water, resource management and consumption, the environmental aspects of development, agriculture and food production.

It originally published as a special issue of *Water International.*

Chittaranjan Ray serves as the Director of Nebraska Water Center at Robert B. Daugherty Water for Food Institute at the University of Nebraska, Lincoln, USA, and is a Professor in the Department of Civil Engineering. Besides his leadership role in water at the university, his research focuses on water quality impacts from agriculture.

David McInnes is the Principal of DMci Strategies, where he is a Strategic Advisor, Speaker and Facilitator on change and opportunity facing the global food system. He is also a Senior Fellow at Canada 2020, an independent, progressive think-tank, where he leads the Canada Food Brand project. Dr McInnes is also a Contributing Editor of The Economist Intelligence Unit, and an Advisor to the Delegation of Canada for the Inter-American Institute for Cooperation on Agriculture.

Matthew R. Sanderson is the Randall C. Hill Distinguished Professor of Sociology, Anthropology and Social Work at Kansas State University, Manhattan, USA. His research focuses on the demographic and environmental aspects of development in global political-economic context.

Routledge Special Issues on Water Policy and Governance

Edited by
Cecilia Tortajada (IJWRD) – Institute of Water Policy, Lee Kuan Yew School of Public Policy, NUS, Singapore
James Nickum (WI) – International Water Resources Association, France

Most of the world's water problems, and their solutions, are directly related to policies and governance, both specific to water and in general. Two of the world's leading journals in this area, the *International Journal of Water Resources Development* and *Water International* (the official journal of the International Water Resources Association), contribute to this special issues series, aimed at disseminating new knowledge on the policy and governance of water resources to a very broad and diverse readership all over the world. The series should be of direct interest to all policy makers, professionals and lay readers concerned with obtaining the latest perspectives on addressing the world's many water issues.

OECD Principles on Water Governance
From Policy Standards to Practice
Edited by Aziza Akhmouch, Delphine Clavreul, Sarah Hendry, Sharon Megdal, James Nickum, Francisco Nunes-Correia and Andrew Ross

Urban Resilience to Droughts and Floods
The Role of Policies and Governance
Edited by Cecilia Tortajada, James Horne and Larry Harrington

Politics and Policies for Water Resources Management in India
Edited by M. Dinesh Kumar

Rural–Urban Water Struggles
Urbanizing Hydrosocial Territories and Evolving Connections, Discourses and Identities
Edited by Lena Holmes, Rutgerd Boelens, Leila M. Harris and Gert Jan Veldwisch

Legal Perspectives on Bridging Science and Policy
Edited by Mara Tignino, Raya Marina Stephan, Renée Martin-Nagle and Owen McIntyre

Virtual Water
Implications for Agriculture and Trade
Edited by Chittaranjan Ray, David McInnes and Matthew R. Sanderson

For more information about this series, please visit: https://www.routledge.com/series/WATER

Virtual Water

Implications for Agriculture and Trade

Edited by

Chittaranjan Ray, David McInnes and Matthew R. Sanderson

Routledge
Taylor & Francis Group
LONDON AND NEW YORK

IWRA

First published 2020
by Routledge
2 Park Square, Milton Park, Abingdon, Oxon, OX14 4RN

and by Routledge
52 Vanderbilt Avenue, New York, NY 10017

Routledge is an imprint of the Taylor & Francis Group, an informa business

British Library Cataloguing-in-Publication Data
A catalogue record for this book is available from the British Library

ISBN13: 978-0-367-40807-7

Typeset in Minion Pro
by codeMantra

Publisher's Note
The publisher accepts responsibility for any inconsistencies that may have arisen during the conversion of this book from journal articles to book chapters, namely the inclusion of journal terminology.

Contents

Citation Information

The chapters in this book were originally published in *Water International*, volume 43, issue 6 (August 2018). When citing this material, please use the original page numbering for each article, as follows:

Introduction
Virtual water: its implications on agriculture and trade
Chittaranjan Ray, David McInnes and Matthew Sanderson
Water International, volume 43, issue 6 (August 2018) pp. 717–730

Chapter 1
The water footprint of the EU: quantification, sustainability and relevance
Davy Vanham
Water International, volume 43, issue 6 (August 2018) pp. 731–745

Chapter 2
The exposure of a fresh fruit and vegetable supply chain to global water-related risks
Tim Hess and Chloe Sutcliffe
Water International, volume 43, issue 6 (August 2018) pp. 746–761

Chapter 3
Advising Morocco: adopting recommendations of a water footprint assessment would increase risk and impair food security for the country and its farmers
Dennis Wichelns
Water International, volume 43, issue 6 (August 2018) pp. 762–784

Chapter 4
Future crop yields and water productivity changes for Nebraska rainfed and irrigated crops
Yaqiong Lu, Xianyu Yang and Lara Kueppers
Water International, volume 43, issue 6 (August 2018) pp. 785–795

Chapter 5
Can Sub-Saharan Africa feed itself? The role of irrigation development in the region's drylands for food security
Hua Xie, Nicostrato Perez, Weston Anderson, Claudia Ringler and Liangzhi You
Water International, volume 43, issue 6 (August 2018) pp. 796–814

Chapter 6
Sustainability of aquifers supporting irrigated agriculture: a case study of the High Plains aquifer in Kansas
James J. Butler Jr., Donald O. Whittemore, B. Brownie Wilson and Geoffrey C. Bohling
Water International, volume 43, issue 6 (August 2018) pp. 815–828

Chapter 7
Irrigation variability and climate change affect derived distributions of simulated water recharge and nitrate leaching
Timothy R. Green and Saseendran S. Anapalli
Water International, volume 43, issue 6 (August 2018) pp. 829–845

Chapter 8
The water footprint challenge for water resources management in Chilean arid zones
Pablo Álvarez
Water International, volume 43, issue 6 (August 2018) pp. 846–859

Chapter 9
The effect of diet changes and food loss reduction in reducing the water footprint of an average American
Mesfin M. Mekonnen and Julian Fulton
Water International, volume 43, issue 6 (August 2018) pp. 860–870

Chapter 10
Water footprint for Korean rice products and virtual water trade in a water-energy-food nexus
Sang-Hyun Lee, Jin-Yong Choi, Seung-Hwan Yoo and Rabi H. Mohtar
Water International, volume 43, issue 6 (August 2018) pp. 871–886

Chapter 11
Water footprint of beef production on Texas High Plains pasture
Charles P. West and Lisa L. Baxter
Water International, volume 43, issue 6 (August 2018) pp. 887–891

Chapter 12
Tradeoffs in the water-energy-food nexus in the urbanizing Asia-Pacific region
Makoto Taniguchi, Naoki Masuhara and Shun Teramoto
Water International, volume 43, issue 6 (August 2018) pp. 892–903

Contributors

Pablo Álvarez is an Agronomist from the University of Chile, Santiago, Chile. He was the Director of the Department of Agronomy of the University of La Serena (1998–2001), Chile and the Director of CEAZA (2006–2008). His field of expertise includes irrigation, agricultural hydrology and water resource management under restrictive conditions.

Saseendran S. Anapalli is a Research Soil Scientist with the USDA Agricultural Research Service (ARS) in the Center for Agricultural Resources Research.

Weston Anderson is a Postdoctoral Research Scientist in the Department of Earth and Environmental Sciences at Columbia University, New York City, USA. His fields of interest are drought, hydroclimate and food security.

Lisa L. Baxter is an Assistant Professor and Extension Forage Agronomist in the Department of Crop and Soil Science at the University of Georgia, Athens, USA.

Geoffrey C. Bohling is an Associate Scientist at the Kansas Geological Survey at the University of Kansas, Lawrence, USA.

James J. Butler Jr. is a Senior Scientist and Geohydrology Section Chief of the Kansas Geological Survey at the University of Kansas, Lawrence, USA.

Jin-Yong Choi is in the Department of Rural Systems Engineering and Research Institute for Agriculture & Life Sciences at Seoul National University, Republic of Korea.

Julian Fulton is an Assistant Professor in the Department of Environmental Studies at California State University, Sacramento, USA.

Timothy R. Green is a Research Scientist with the USDA Agricultural Research Service (ARS) in the Center for Agricultural Resources Research, and a Faculty Affiliate in the Department of Civil and Environmental Engineering at Colorado State University, Fort Collins, USA.

Tim Hess is a Professor of Water and Food Systems at Cranfield University, UK. He has over 35 years' experience in research and consultancy in the application of hydrology in managed rural environments, and particularly irrigated agriculture in the UK and overseas. This has been applied to sustainable food supply chains, working with national and international organisations and major UK retailers. He is an interdisciplinary scientist with a particular interest in the application of ecosystem services approaches to water management issues.

Lara Kueppers is an Assistant Professor in the Energy and Resources Group at the University of California Berkeley, USA, with a Faculty Scientist appointment at Berkeley Lab. She is an interdisciplinary environmental scientist, whose research focuses on ecological responses and feedbacks to climate change.

Sang-Hyun Lee is an Associate of Geophysical Fluid Dynamics Institute in the Department of Biological and Agricultural Engineering at Texas A&M University, College Station, USA.

Yaqiong Lu is a Guest Postdoc Fellow in the Climate and Ecosystem Sciences Division at Lawrence Berkeley National Laboratory from the University of California, Merced, USA.

Naoki Masuhara is a Project Researcher at the Research Institute for Humanity and Nature (RIHN), Japan. His major fields of interests are local environment administration and energy policy.

David McInnes is the Principal of DMci Strategies, where he is a Strategic Advisor, Speaker and Facilitator on change and opportunity facing the global food system. He is also a Senior Fellow at Canada 2020, an independent, progressive think-tank, where he leads the Canada Food Brand project. Dr McInnes is also a Contributing Editor of The Economist Intelligence Unit, and an Advisor to the Delegation of Canada for the Inter-American Institute for Cooperation on Agriculture.

Mesfin M. Mekonnen is a Professor in the Water for Food Institute at the University of Nebraska, Lincoln, USA. Mekonnen is interested in understanding the interactions of human and natural systems in determining the sustainability of freshwater resources. His specialties include water footprint assessment, water scarcity, water management, crop water use modeling, spatial modeling and water for energy.

Rabi H. Mohtar is a TEES Research Professor at Texas A&M University, College Station, USA. He is the Founding Director of Qatar Environment and Energy Research Institute (QEERI), a Member of Qatar Foundation Research and Development and the Founding Director of Strategic Projects at Qatar Foundation Research and Development. His primary research priority is the development of a framework to quantify the interlinkages of the water-energy-food nexus that is constrained by climate change and social, political and technological pressures.

Nicostrato Perez is a Senior Scientist in the Environment and Production Technology Division of the International Food Policy Research Institute. He has more than 30 years of research experience, working in various capacities in national and international agricultural research centers. His research interests are water and food security, adaptation to climate change, sustainable intensification and economy-environment policy modeling.

Chittaranjan Ray serves as the Director of Nebraska Water Center at Robert B. Daugherty Water for Food Institute at the University of Nebraska, Lincoln, USA, and is a Professor in the Department of Civil Engineering. Besides his leadership role in water at the university, his research focuses on water quality impacts from agriculture.

Claudia Ringler is the Deputy Division Director of the International Food Policy Research Institute's Environment and Production Technology Division. Her research interests are water resources management – in particular, river basin modeling for policy analysis and agricultural, and natural resource policy focused at sustainable agricultural productivity growth.

Matthew R. Sanderson is the Randall C. Hill Distinguished Professor of Sociology, Anthropology and Social Work at Kansas State University, Manhattan, USA. His research focuses on the demographic and environmental aspects of development in global political-economic context.

Chloe Sutcliffe is a Research Fellow in Water and Food System Resilience at Cranfield University, UK. She is currently working on a three-year project funded through the Global Food Security Programme, which aims to identify how and where the UK's fresh fruit and vegetable system is exposed to water-related risk and determine how these risks may change over time.

Makoto Taniguchi is a Hydrogeologist at the Research Institute for Humanity and Nature (RIHN), Japan. He moved to RIHN in 2003, and became a Full Professor of hydrogeology in 2009. He is also the President of the Japanese Association of Groundwater Hydrology (2015–present). He has worked on groundwater projects around the world, in particular Asia.

Shun Teramoto is a specially appointed specialist at the Research Institute for Humanity and Nature (RIHN), Japan.

Davy Vanham is a Scientist in Water Management at the Joint Research Centre of the European Commission in Ispra (IT). His research focuses on the interdisciplinary management of water resources and its uses, in order to find sustainable solutions to the world's water-related challenges. He has 20 years of working experience in IWRM both in the private and public sectors.

Charles P. West occupies the Thornton Distinguished Chair in Plant and Soil Science, with specialization in forage systems research at Texas Tech University, Lubbock, USA. Dr West provides administrative leadership of the Texas Alliance for Water Conservation, a consortium of agricultural producers, water districts, government agencies and researchers to refine and transfer irrigation technology to make efficient use of water supplies. He is also the Director of the College of Agricultural Sciences and Natural Resources Water Center.

Donald O. Whittemore is a Senior Scientific Fellow in Environmental Geochemistry at the Kansas Geological Survey at the University of Kansas, Lawrence, USA.

Dennis Wichelns is an Agricultural and Natural Resource Economist and is a Senior Research Fellow with the Stockholm Environment Institute. Wichelns has also served as a Senior Research Fellow and Deputy Director for research with the International Water Management Institute, Chief Economist with the California Water Institute and Expert Witness in the U.S. Supreme Court. He is the Editor-in-Chief of two journals: "Agricultural Water Management" and "Water Resources & Rural Development."

B. Brownie Wilson is a Research Assistant in Geohydrology and GIS/Support Services Manager at the Kansas Geological Survey at the University of Kansas, Lawrence, USA.

Hua Xie is a Research Fellow in the Environment and Production Technology Division at the International Food Policy Research Institute, Washington, D.C., USA. His specialty area is water resources and environmental system analysis. Topics of his research at IFPRI include irrigation investment analysis, global water quality modeling, large-scale hydrologic and crop simulation, climate change impact assessment and sustainable land management.

Xianyu Yang is in the School of Atmospheric Sciences at the Chengdu University of Information Technology, China.

Seung-Hwan Yoo is in the Department of Rural and Biosystems Engineering at Chonnam National University, Gwangju, Republic of Korea.

Liangzhi You is a Senior Scientist at the International Food Policy Research Institute, where he conducts research on agricultural science and technology policy.

Acknowledgments

This book is derived from a workshop on "Virtual Water: Its Implications on Agriculture and Trade", which took place in September 2016 at the University of Nebraska-Lincoln, USA. The workshop was sponsored by the OECD Co-operative Research Programme: Biological Resource Management for Sustainable Agricultural Systems, whose financial support made it possible for some of the invited speakers to participate in the workshop.

The opinions expressed and arguments employed in this publication are the sole responsibility of the authors and do not necessarily reflect those of the OECD or of the governments of its Member countries.

EDITORS' INTRODUCTION

Virtual water: its implications on agriculture and trade

Chittaranjan Ray, David McInnes and Matthew R. Sanderson

Water is the basis of life. And, unlike various forms of energy generation, water cannot simply be created where it is needed. Freshwater is remarkably scarce, comprising just 2.5% of the global water supply. Although this water at an annual level may be judged as sufficient for human use, spatio-temporal variations in the availability of freshwater over the globe is a challenge. Growing populations and rising levels of affluence mean increased competition for water, raising vital questions of equity, access, and social justice at the global scale. Water is thus a lens through which to examine an array of vital issues facing humanity and the planet: human and animal health; food production; environmental management; resource consumption; climate change adaptation and mitigation; economic development, trade and competitiveness; and ethics and consumer trust – to name but a few.

The articles here, arising from a workshop supported by the OECD Co-operative Research Programme: Biological Resources Management for Sustainable Agricultural Systems, on virtual water, agriculture, and trade at the University of Nebraska-Lincoln in September 2016, consider questions of gaps in knowledge, why sustainability matters, and the policy implications of virtual water trade.

What we don't know can hurt us

A pervasive theme throughout the articles herein, and the workshop that initiated them, is the general lack of knowledge of the use of water in producing and consuming food. For instance:

- Consumers are generally unaware, when they purchase imported goods, that they are effectively outsourcing water depletion in the form of crop or food demands in another country or region. This can be described as 'off-shoring environmental burdens' (Hess and Sutcliffe, this issue).
- Agricultural regions suffering from droughts invite public scrutiny about producers' water practices (e.g. irrigation) particularly when consumers in nearby cities are subjected to water advisories or supply cutbacks. More broadly, managing water issues is about maintaining consumer trust among citizens who are often far-removed from the source and use of that water (at the farm/ranch level).

- The lack of good data, computing tools, and interpretation techniques (in the context of future climatic scenarios and impact on water supply) is a major issue in terms of how water is managed now and in the future. For example, not all irrigation wells in the High Plains/Ogallala Aquifer (HPOA) are metered to provide data on pumpage. Even when adequate data are available, improved water footprint calculation methods are required to properly understand and manage stressed water resources.
- As water stresses increase, we need better transparency of policies that incentivize or determine water use in agriculture. Water is not generally priced in terms of its cost of depletion. Various subsidies hide the true cost of water which causes market-distorting behaviors with negative environmental effects.
- Also unclear is whether food producers and the food system in general (i.e., processors and retailers) fully understand the long-term implications of unsustainable water use from aquifers. Put another way, market signals may not be incentivizing the right water use practices. That said, food supply chains are becoming far more focused on sustainable sourcing, and demonstrating that to consumers. This shift could be a game-changer to make current practices to be more sustainable.

Understanding virtual water and the water footprint

Virtual water is the amount of water 'embedded' in a product (Allan, 2011). As agricultural products are sold and traded, the water that is used to produce them (natural rainfall or irrigation) is also essentially traded. Identifying the amount of virtual water embedded in a product has implications for water management, practice, and policy. For instance, with irrigation, many arid regions around the world are major producers of our food and feed, and an important source of agricultural exports. Yet, increasing climate stress and competition for scarce water in these regions raise significant questions for stakeholders in an integrated, globalizing food system: Where should food be produced? How best to use limited water resources? What information should be available to consumers to assist them in making more informed decisions?

The 'water footprint' (Hoekstra & Chapagain, 2008; Hoekstra & Mekonnen, 2012) has been developed as a tool, especially by the Water Footprint Network, to measure virtual water use over the entire supply chain from production to delivery to the consumer, and including pollution produced in the process. It refers to the amount of fresh water it takes to produce a product (whether a food or material good), to bring it to the consumer, and in dealing with the pollution created by the product over the supply chain. Vanham's article here describes two methodologies to conduct water footprint assessments – one developed within the Integrated Water Resources Management (IWRM) framework, and one associated with work in the Life Cycle Analysis framework. The author shows that there are non-trivial differences between the methodologies, but also complementarities.

The water footprint is typically divided into three sub-categories: blue, green, and grey (see Figure 1). Blue water refers to the consumption of surface and ground waters (by irrigation) along the supply chain of a product or from growing a crop to bringing it

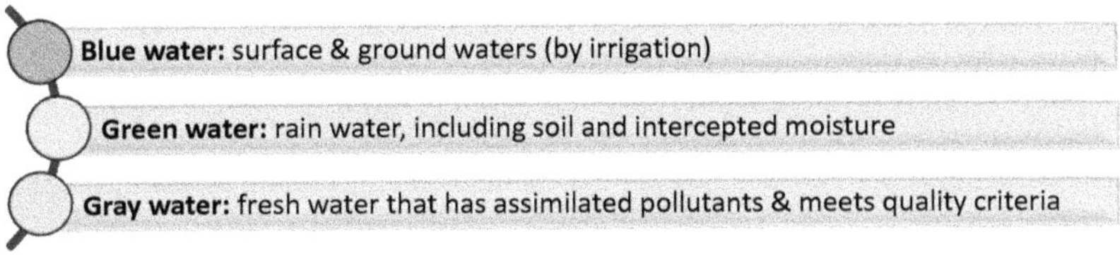

Figure 1. Water footprint categories based on consumption.

to market. Green water relates to the consumption of rainwater (before it becomes runoff). This includes the moisture available to plants from soil and intercepted moisture in the plant canopy or on soil surfaces. The term 'consumption' used in blue and green waters refers to incorporation into the product, evaporation from soil or transpiration from plants, deep percolation to ground water, or movement of surface water from one catchment to another. Grey water is the amount of fresh water that will be required to assimilate the pollutant loads to meet water quality criteria as set by local regulations (i.e., dilution requirements for pollution).

There is often disagreement over the procedures adopted for water footprint calculations. For example, the water footprint of a kilogram of wheat or kilogram of beef often represents an average when expressed as a global number. Averages can conceal important qualitative differences among areas of production, and these figures can often vary from place to place. When it comes to distinguishing blue water (often the more costly component) from green water, discrepancies in methods can be important. For example, the water footprint of a kilogram of beef can have different blue or green water footprints if the cattle are raised in a ranch versus in a confined animal feeding operation (CAFO). The animal feed for the CAFO cattle may be produced using blue water or green water, or some combination. For cattle raised on rangelands or ranches, a major part of the animal's diet comes from natural grass, which is dependent on green water. In the mid-western USA beef cattle production system in the rangelands, some grain is used to fatten the animal before selling into the market, but this method of cattle production would have a lower blue water footprint compared to fully grain-fed beef in these areas. In many parts of the world, the beef cattle raised in the range land or ranches are sold directly in the market rather than fattening them in CAFOs. In such cases, the total water footprint may be high due to low feed to product conversion ratio (i.e., more grass is needed for the same mass of beef than corn or other feed). When considering meat from swine or chicken, most feeding and raising of these animals occurs in CAFOs. Producing animal protein has a definable water footprint at a given location and the blue and green water component will vary. In short, making clear comparisons and understanding the trade-offs of water use is not always clear, can make public dialogues confusing and contentious, and can provide variabilities in understanding environmental impacts. Thus, while it can be useful for decision-making, using the water footprint is not without difficulties. For instance, Wichelns (this issue) argues that water footprint ratios have limitation because they do not consider the conditions in which crops are grown. Going forward, it will be important to clearly understand the advantages and the limitations of water footprint, and virtual water measures, especially if they are used to inform policy and practice.

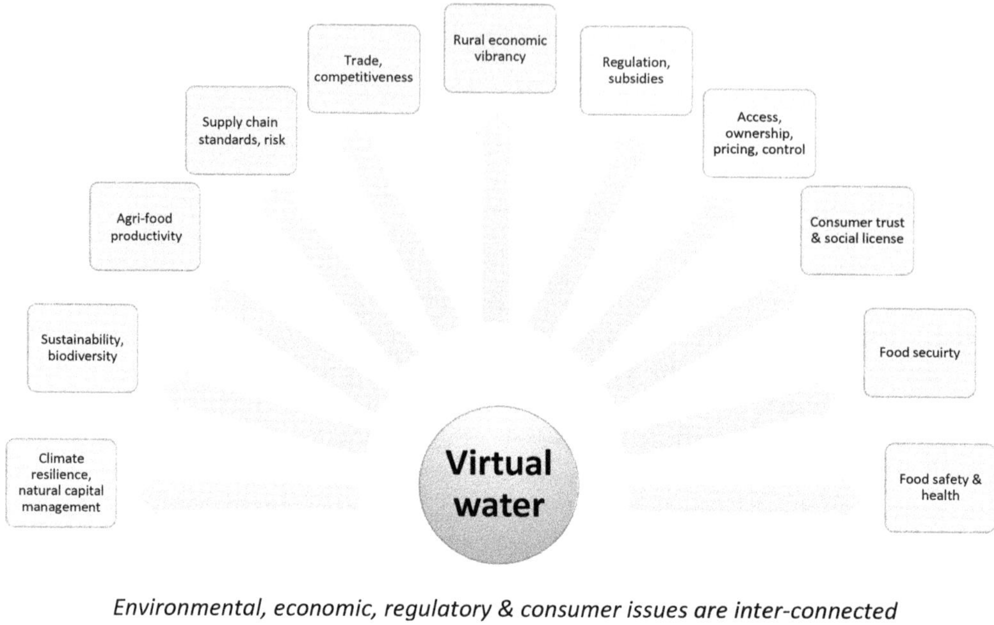

Figure 2. Water is a lens to examine key issues facing agriculture, people, and the planet.

How we grow, process, trade and consume our food is intimately linked to how we use and manage water. Below, we elaborate on insights emerging from the articles in this issue.

Perspectives on blue water

Many of the contributions here focus on one of the most productive growing regions on Earth, the American Great Plains, but we have included a number of insightful international cases. Combined, this offered a lens (see Figure 2) to explore a variety of major issues about water used in modern agriculture and the global food system – a number of which are elaborated upon below.

Food production and food security

In the HPOA region of the American Midwest, large-scale production of corn, soybean, wheat, cotton, and sorghum rely heavily on locally available groundwater (Martson, Konar, Cai, & Troy, 2015). However, in many parts of the region, irrigation pumping exceeds the recharge rates (Scanlon et al., 2012), resulting in groundwater depletion. In southern part of the HPOA (particularly in Texas, Oklahoma, and western Kansas), groundwater depletion has been so severe that existing irrigation systems are unable to fully irrigate fields. As Konar et al. show, other large aquifers such as the Central Valley and the Mississippi Embayment are also experiencing significant overdraft and long-term sustainability concerns. Their paper raises vital questions about temporal and spatial trade-offs. Clearly, consumption and production today comes at the expense of future use. This is apparent for the HPOA, which is essentially nonrenewable in its southern and central regions owing to minimal recharge and high evaporation rates. Tracing virtual water through trade flows can reveal where – or who – is consuming the water embedded in the agricultural products. These analyses are raising important, and difficult, questions for policy in an increasingly inter-connected global food system.

One key issue is about global accountability. While considerable attention is devoted to the smooth functioning of global supply chains, which results in a relatively reliable and safe food supply for consumers, less consideration appears to be given to deciding what institutions and capacity are needed to properly manage, or govern, water consumption at requisite scales. Improving transparency of water allocation and management practices may help address this challenge.

Future food security is also made more challenging because depletion is accompanied by competition for, and increased scrutiny of, the water resource (World Water Assessment Programme [WWAP], 2012). Agriculture competes for water with other economic interests. Oil exploration and processing rely on ground water from the HPOA in Southwest Kansas, West Texas, New Mexico and Oklahoma. Competition for water creates a contentious debate over who should have access to it. The situation also puts a greater spotlight on big users of water, including farming practices. For instance, in the Great Plains of the US, many farmers have been producing consecutive corn crops over the last decade or so, a more intensively irrigated and fertilized commodity than other common crops produced in the region. Producers respond to customer demand signals for their crops and to government signals (subsidies). The push to increase yields is prompting more interest in sustainability practices, such as how nitrogen fertilizer is used and managed (which can affect water quality) and in prompting watershed-wide analyses of irrigation practices.

These issues will likely become even more important in the context of a changing climate. The article by Lu et al., for example, finds that in Nebraska, irrigated crops are less sensitive to changing climate than dryland crops, but irrigated crops have lower water productivity. Among their findings, they suggest that expanding irrigated winter wheat acreages could maintain high yields while conserving more water. Similarly, the article by West and Baxter reports that the legume lucerne (alfalfa, *Medicago sativa*) could be introduced to lower the water footprint of the stocker phase of beef production in the U.S. High Plains. Informed choices about crop mixes and rotations are crucial for improving the sustainability of agri-food systems in changing climates.

The case has also been made for expanding irrigation investment in certain regions. As Xie et al. argue in this volume, irrigation may help address Sub-Saharan Africa's food security and nutrition situation and reduce food import dependency. Small-scale irrigation could substantially benefit the production of vegetables and groundnuts. Such activity has a multiplier effect, such as on providing employment and trade opportunities. However, the UN Sustainable Development Goals (SDGs) offer guidance here. Finding the optimum pathway to eliminate hunger and malnutrition (Goal 2) must include ensuring sustainable water withdrawals (Goal 6) and pursuing sustainable use of natural resources that takes into account the situation in developing countries (Goal 12).

Environmental sustainability

The environmental sustainability of water used in agriculture is complex. Understanding impacts on primary aquifers is paramount. As sources of irrigation water, aquifers can become stressed when human demand (pumpage) outstrips natural recharge rates and captures water that would otherwise feed surface flows (Bredehoeft, 2002; Gleeson, Wada, Bierkens, & van Beek, 2012). Few argue that aquifers can be practically restored to the original (pre-pumping) state unless agriculture is completely or significantly curtailed – the politics of which are not feasible. In such situations, the focus shifts to how to do 'no more harm.' This largely reflects the situation in HPOA today, although the magnitude of the problem varies across the region (i.e., north to south). For example, the southern region of the HPOA (western Kansas, and the panhandle regions of Oklahoma and Texas) has seen significant declines in water levels compared to the pre-development era (Haacker, Kendall, & Hyndman, 2016). Some areas have seen water levels lowered more than 150 feet (45 metres) and other areas are no longer irrigable (McGuire, 2014). At this rate of consumption, and with current cropping systems in place, even more of the southern and central parts of the HPOA are expected to face questions about the longevity of the aquifer for irrigated crop production.

It is apparent that good data are required so that modeling and analytics can help us respond to the challenges of water trade. As Butler et al.'s paper shows, water use and water level data can be used to increase the effectiveness of forecasting models. Kansas data, in this sense, are rare in the detail they provide analysts, but when such data are available, the insights are exceedingly useful for understanding that policies needed to extend the useful life of the resource. Butler et al. find that 'practically feasible pumping reductions' of 10–20% would have large impacts on rates of water decline, and that slightly larger reductions (20–30%) could even lead to near-stable water levels. Yet, there remains much room for improvement in the area of water data collection and availability. For example, while flow meters are required in western Kansas, many irrigation wells in the HPOA are not yet metered to provide data on pumping, which is a more direct and reliable measurement of the volume of groundwater extracted. The absence of complete data on water consumption is a clear challenge for scientists' capacity to model and assess impacts. In the context of future climate change scenarios, an increase in the temperature in the summer growing season is likely to enhance evapotranspiration, which will increase the demand for irrigation. Changing climates and the persistence of droughts (such as the 2012 drought in the Great Plains region) puts increasing pressure on the aquifer. While the U.S. Geological Survey has also had a monitoring program for HPOA water levels since the mid-1980s, the need for good data and better water footprint calculation methods becomes more important as this resource becomes more stressed.

Water pricing

Pricing water is directly relevant to the economics of sustainability. It is controversial. Politicians, citizens and the agri-food sector need to confront trade-offs associated with the reliance on government subsidies to produce feed, food, fiber and fuel from agricultural crops. Subsidies may be well-intentioned but they can hide the true cost of water and encourage unsustainable agricultural practices. In Nebraska, for example, price supports and subsidies for corn have changed crop rotations as producers seek consecutive high yields. What seemed at one time like an endless bounty of groundwater, combined with relatively inexpensive energy to pump it, has created significant momentum to continuously grow the crops that show the greatest short-term profit. In the U.S., federal government subsidies, including the corn ethanol subsidy, support this practice.

While farmers are generally motivated for cost reasons to reduce nitrogen fertilization use and to maintain healthy soils, consecutive cropping of corn over the past decades in this region has led to significant input of nitrogen fertilizers (Exner, Hirsh, & Spalding, 2014; Juntakut, Snow, Haacker, & Ray, submitted). This has led to high nitrate levels in groundwater which has impacted many communities, challenging them in providing water that meets health standards for human consumption. Fertilizer impacts also vary with soil type. In the Midwestern corn belt, sandy soils tend to facilitate groundwater contamination. There are also other complications with high nitrates in groundwater. Green and Anapalli provide distributions of water percolation and nitrate leaching based on irrigation variability and projected climate change. Groundwater nitrate contamination can also increase the risk of further contamination from naturally-occurring toxins. Denitrification can enhance the mobilization of uranium naturally present in the subsurface (Nolan & Weber, 2015). Notwithstanding the fact that agriculture is not the only source of water contamination, local communities are still left with the task of ensuring drinking water quality and adequacy for its citizens. Grand Island and McCook are among the Nebraska communities that have installed treatment plants for uranium.

In essence, citizens pay for water three times. As Vanham describes, taxpayers pay for the farm subsidy that encourages corn production; consumers pay a hidden price when purchasing food (as water has some cost of food production or processing as an input cost); and residents pay for any municipal/regional cleanup or rehabilitation cost of the groundwater from production practices. While there is widespread pressure by cities to require the farmers or farmers associations to pay for cost of treating water contaminated with nitrate or pesticides, there is no precedence in the United States to pay for this cost. In a recent court case in Iowa, the City of Des Moines filed a lawsuit against counties upstream from its storage reservoir that provides water to its residents. The claim in the lawsuit was to recover cost of treatment; however, this lawsuit was thrown out by the Iowa Supreme Court, which ruled that it was a legislative matter, not a judicial matter.[1] There are many small towns in Nebraska, Iowa, Illinois and other Midwestern states where the farmers indirectly pay for cost of treatment because they live in those towns, and the town increases the cost of supplied water and raises property taxes to pay for shortfalls in state and federal funding of treatment plants.

Water pricing can be used to manage water as a scarce resource and achieve improved productivity. Engaging producers on the economics of water is a matter of self-interest and societal benefit – and should be seen as such. Water productivity is a measure of the amount of water used per unit of agricultural product: the higher the water use relative to yield, the lower the efficiency. However, a crop simulation model has shown that farmers tend to apply more irrigation water than what is needed (Carr, Yang, & Ray, 2016). Use of variable rate irrigation to reduce overwatering in different soil types, and installation of variable drive pumps to reduce energy demand (generated from electricity, diesel or natural gas), are a few of the features available to reduce overwatering. The ultimate objective – which is in the best interests of farmers and society – is more sustainable water use: less blue water used per unit of production. Widely adopting this key measure of efficiency appears to be an essential farm management and sustainability tool. However, additional economic analysis is warranted, including how to promote its widespread adoption.

Ownership

Water ownership is central to addressing sustainability. In Chile, for example, Alvarez describes how water law fails to promote sustainable water use by allocating water rights according to basin water supply rather than crop irrigation demand. Alvarez's research is a reminder that how we manage water is, essentially, a governance issue. In the U.S. context, water rights and ownership vary significantly among the states. This invites different sustainability options, including the role government can play depending on state or individual rights of water. For example, Peck (2007) states that Nebraska uses a hybrid approach to managing the ground water rights, which includes the 'Reasonable Use Doctrine' and the 'Correlative Rights Doctrine.' The right to use comes from the ownership of land overlying the water. If the site is within a groundwater management area, a permit is needed. All well owners are required to register their wells. Statutes also regulate the location of wells with respect to rivers/streams or other wells and the reasonable use clause is employed where the owner cannot use more than a reasonable quantity and it must be shared among others. On the other hand, Texas employs the 'Rule of Capture,' which allows landowners to extract all the water beneath the land and do whatever they want with it. As the landowner is not liable to the neighbor, this has spurred severe aquifer drawdown. However, local regulations are slowly becoming more stringent across the Central and Southern HPOA. In the case of Texas, there are eleven ground water conservation districts (GCDs) in the HPOA region of Texas and each still recognizes landowners as groundwater owners. However, GCDs are required to work with the Texas Water Development Board to enumerate 'desired future conditions,' and to submit a water plan to meet those conditions. These water plans are evaluated using a groundwater model (Deeds & Jigmond, 2015). While restrictions primarily affect cities and their ability to use state funding for new water supply infrastructure, agricultural management is increasingly involved in conservation measures. In 2015, the High Plains Underground Water Conservation District No. 1 implemented metering requirements for irrigation wells, and set a maximum irrigation limit of 18 inches (450 mm) per year for double cropping.

Policy implications

While the workshop was not intended to catalogue every issue relating to water, the dialogue uncovered several pertinent policy issues.

Global trade and agricultural subsidies

Some 70% of the world's fresh water from rivers, lakes, and aquifers (blue water) is used to produce the food for humanity. A great deal of this water is exported in the form of virtual water (Hoekstra & Mekonnen, 2012). Of critical importance is the extent to which this volume of trade has disproportionate negative impacts on the world's water resources. Hess and Sutcliffe's article raises this issue in the context of the United Kingdom's reliance on other countries, such as Spain, to supply many of its fruits and vegetables (a matter that could equally apply to many other countries). As a semi-arid country, Spain is facing severe water stress so its food exports are, effectively, allowing Britain and other nations to 'off-shore' their environmental burdens.

This situation is sustained in part because of the existence of government subsidies. To discuss trade and ecosystem impacts, one must consider the role that agricultural subsidies play in encouraging global agricultural production and groundwater depletion, water quality deterioration and biodiversity impacts. However, subsidies have not drawn a great deal of attention or concern for their impact on the environment; there has not been a World Trade Organization challenge yet against a country based on how it subsidizes water for on-farm use (see Garrido, 2017).

Subsidies exist in many forms. Across the OECD, subsidies (total producer support) have been calculated to be nearly US$240 billion (2014 data) (Canadian Agri-Food Policy Institute [CAPI], 2017). Not all subsidies are necessarily bad; it is the unintended consequences and degree of this impact that is the issue. For instance, while subsidized pricing for fertilizer in India has helped poor farmers in that country, highly subsidized and basically free power use to pump water has not helped the agricultural sector of India. Over-pumping has caused severe groundwater depletion in many areas, particularly the state of Punjab. In 2002, agricultural pumping accounted for more than 31% of total power use in India (Centre for Monitoring Indian Economy [CMIE], 2002).

The issue is difficult to comprehend given that reporting on subsidies has been changing. The 2014 US Farm Bill abolished direct payment based on land ownership, and instead farmers can get subsidies for insurance and for other needs (Polzkill et al., 2017). However, the impact of subsidy practices still masks understanding environmental impacts. The World Trade Organization reporting system, for instance, tends to under-report certain practices that encourage unsustainable practices, including excessive drawdown of aquifers worldwide (Canadian Agri-Food Policy Institute [CAPI], 2017). As demand for food rises and pressure to use water increases, the lack of transparency of these forms of subsidy will only challenge policy-makers further about clearly understanding the causes and effects of their decisions. Meanwhile, the depletion of natural capital will largely continue.

Supply chain risk and resilience

Global processors and retailers depend on reliable supply chains to ensure consumers have safe, reliable and quality products from which to choose every day. These companies are monitoring a variety of supply chain risks, including dependable supply of literally thousands of ingredients and products. Climate change and extreme weather challenges global producers and suppliers, including, as noted earlier, the sourcing of fresh produce from arid growing areas that are highly vulnerable to droughts. Increasingly, these large companies expect producers and supply chains to conform to new sustainability protocols. The action is motivated by demonstrating 'corporate social responsibility' and good corporate governance. Security of supply chains and minimizing reputational risk (i.e., the risk of being accused of not sourcing sustainably) are powerful motivators to change unsustainable practices.

Producers, input suppliers, and processors are also creating bottom-up initiatives to improve crop resilience and sustainability, such as sustainable soy. There are a myriad of initiatives, such as McDonald's Restaurants' participation in a global sustainable beef roundtable that also includes ranchers and many others. Often, major environmental groups are part of these efforts, such as the World Wildlife Fund. A new dynamic is occurring that is linking responsible sourcing with environmental, health, ethical and other objectives. A full assessment is required to discern how these initiatives have implications for groundwater depletion, but the point is that many major supply chains are being responsive to environmental risks.

Technology adoption is also vital to respond to the water challenge and improve sustainability. Technology is key to precision agriculture (and 'precision irrigation'), which will help to ensure that plants get the required nutrients and water at the right time and in the right amount. A global effort is underway to improve fertilizer use (known as the 4R program for 'right source, right rate, right time, and right place'), which is helping to achieve this positive outcome. Farmers are generally motivated to improve their practices as they want to ensure yield and quality, such as using soil moisture sensors to irrigate their fields (and to maximize the ratio of transpiration to evaporation), but the technology is yet to be adopted by all farmers.

What is largely missing is full cost-accounting of the environmental footprint of food production, or at least its widespread adoption within traditional supply chains. As the world population is expected to reach 10 billion by 2050, the need for food is expected to double. The way in which the food system accounts for the impact of food production will become increasingly salient. Moreover, a global effort is underway to promote a new approach, such as with the UN's Sustainable Development Goals. One of its 17 major goals speaks to sustainable production and consumption. Other goals include addressing water use, agricultural production and the environment. At the time of the workshop, the SDGs were not given much consideration since this UN initiative had recently come into being (in early 2016). However, going forward, the SDGs are expected to have increasing relevance for consumers, companies, investors, and governments and are worthy to understand for their implications on managing agri-food system sustainability.

Consumers, health and trust

A major issue facing the sector as a whole is maintaining consumer trust. Mistrust of food production can result in calls for government regulation, shift dietary preferences, invite bad publicity and undermine corporate reputations – and with social media, criticism can be devastating. The move to healthy diets is based on the belief that animal agriculture contributes to environmental degradation (Tilman & Clark, 2014). The Meatless Monday campaign (a global advocacy effort) calls for consumers to reduce meat consumption and improve the health of people and the planet. The water footprint of meat production is much higher than the water footprints for vegetable-based meals (Mekonnen & Hoekstra, 2012). This rationale is used by advocates, consumers, and the media to affirm their position in favor of healthier diets, for the planet and for individuals. The underlying issue is that consumer food choices are increasingly based on expectations of what is good for the environment. A water footprint assessment can provide additional insights into the consumptive effects of diets, food loss, and food waste implicated in producing those diets. For example, Mekonnen and Fulton's article shows that shifting to vegan and vegetarian diets would provide larger reductions in the consumptive water footprint than simply shifting to more healthy diets in the U.S. But, lowering rates of food loss and food waste would decrease the consumptive water footprint in the U.S. even more.

Maintaining rural/agriculture livelihoods is important for all countries. However, most consumers live in urban or peri-urban areas and demands from these populations for sustainable food productions are becoming more pronounced, particularly in wealthier countries. Urban consumers are disconnected from the farm or ranch but still have certain expectations about how their food needs to be produced (e.g., hormone-free, GMO-free, organic, shade-grown, etc.). It is true that such expectations create new marketing opportunities and product innovation opportunities for food suppliers, but there will continue to be a tension in the system to respond to food wants from an increasingly urban consumer population.

The agri-food sector generally responds to these issues with the call for more consumer education. Certainly, better information is needed. However, most consumers will not likely spend much time considering the distinctions between blue and green waters. While most consumers make food choices based on price, taste and quality, consumers are starting to make value decisions based on sustainability, in many cases as a result of advocacy from environmental non-profit groups. This explains why food retailers provide sustainably-sourced seafood and processors certify themselves as protecting ecosystems. Local food is also preferred in part because consumers feel that the carbon footprint is less than food produced from afar, since the energy inputs for transportation are among the most visible and easily grasped components of the food production system.

Understanding sustainability footprints is complex. Lee et al. describe in this volume the usefulness of the water footprint for rice products and the food-energy-water nexus in Korea. But there are trade-offs in the food-energy-water nexus, especially given rapid urbanization, as Taniguchi et al. discuss in the Asia-Pacific region. Another example is to consider the feed needed for beef cattle production in Nebraska. If the feed is sourced from local cooperatives using Nebraska groundwater for corn irrigation, then the water footprint per kilo of beef is expected to be lower compared to the footprint of the feed corn that is transported from Texas, Oklahoma, or southwestern Kansas, where the evapotranspiration demand is high and more water is used to produce similar amount of feed corn. This is a calculation that consumers would obviously want experts to determine. If it is important to consumers, they would just want to trust the verification of any such claims. Often consumers do not trust self-appointed experts because the people with the greatest stake in the water footprint are the 'experts' hired by food companies, and they tend to find a far lower water footprint than independent analyses (Nestle, 2015).

As sustainability issues intensify, the agri-food sector and regulators will need to decide what are the best certification and verification protocols to deploy in order to maintain societal and customer trust. The level and transparency of that bar can only be expected to rise.

Going forward

Irrigation water will continue to be vitally important for agricultural production, the food supply, trade, and for the economic viability of communities. In the case of groundwater resources of the HPOA, can depletion rates continue for shorter term benefits at the expense of long-term sustainability? The answer to agricultural water management problems is made difficult by how we price the extraction of this precious resource, how we incentivize its draw-down and protect it through agricultural practices, and how we price our food and make it visible or not to the final consumer. Surface water diversions often affect resources that cross political boundaries, such as the Republican and Colorado rivers and the Aral Sea in central Asia. The issue has geopolitical and ethical ramifications for the developed and developing world alike. Wealthier economies import many foods from water-stressed regions. In Africa, countries are leasing large tracts of land to produce food on a contract basis for export. Consumers get the benefit of this production, but at the local cost of water depletion, the diminishing resource for current agriculture and the impacts on water quality from excessive or cumulative nutrient-loading or increased salinity of drinking water. This is a legacy we are seeing unfold today and leaving to future generations tomorrow.

The necessary counter-balance could be a combination of technology and transparency of policies and practices, made possible with better data and analytics, plus significant shifts in attitudes. We may be able to better manage aquifers and the food supply chains that depend on them. But this demands an integrated approach for food systems and communities. Documenting the unforeseen consequences of current practices is a vital step to transforming how we produce our food. The global food system is adopting, certifying and branding a host of sustainably-sourced commodities and foods, from sustainable palm oil and soy to reduce the incidence of rainforest deforestation, to sustainable beef and seafood to protect the food supply and the environment. To effect real change, we may have no choice but to put 'sustainably-sourced aquifers' on this list as well.

Disclosure statement

No potential conflict of interest was reported by the authors.

Note

1. https://www.scribd.com/document/342452360/Ruling-on-Summary-Judgment-00795075xB8DF0.

Acknowledgments

This special issue developed from the workshop on Virtual Water in Agricultural Products: Quantification, Limitations and Trade Policy (Lincoln, Nebraska, USA, 14-16 September 2016), which was sponsored by the OECD Co-operative Research Programme: Biological Resource Management for Sustainable Agricultural Systems whose financial support made it possible for the author to participate in the workshop.

References

Allan, T. (2011). *Virtual water: Tackling the threat to our planet's most precious resource*. London: I.B. Tauris & Co Ltd.

Bredehoeft, J. D. (2002). .The water budget myth revisited: Why hydrogeologists model. *Groundwater*, *40*(4), 340–345.

Canadian Agri-Food Policy Institute (CAPI). (2017) Retrieved from http://www.capi-icpa.ca/pdfs/2017/CAPI_Subsidies_Summary_2017_ENG.pdf

Carr, T., Yang, H., & Ray, C. (2016). Temporal variations of water productivity in irrigated corn: An analysis of factors influencing yield and water use across central Nebraska. *PLoS ONE*, *11* (8), e0161944.

Centre for Monitoring Indian Economy (CMIE). (2002). *Energy 2002*. Mumbai, India: Economic Intelligence Service, Centre for Monitoring Indian Economy.

Deeds, N. E., & Jigmond, M. (2015). *Numerical Model Report for the High Plains Aquifer System Groundwater Availability Model*. Prepared for the Texas Water Development Board, Austin, TX.

Exner, M. E., Hirsh, A. J., & Spalding, R. F. (2014). Nebraska's groundwater legacy: Nitrate contamination beneath irrigated cropland. *Water Resources Research*, *50*, 4474–4489.

Garrido, A. (2017). La importancia del comercio internacional y la globalización para asegurar la seguridad hídrica y alimentaria. Conference paper presented at *Virtual Water in Agricultural Products: Quantification, Limitations and Trade Policy*, Lincoln, NE, USA, 2016 September 14–16.

Gleeson, T., Wada, Y., Bierkens, M. F. P., & van Beek, L. P. H. (2012). Water balance of global aquifers revealed by groundwater footprint. *Nature*, *488*(7410), 197–200.

Haacker, E. M. K., Kendall, A. D., & Hyndman, D. W. (2016). Water level declines in the high plains aquifer: Predevelopment to resource senescence. *Groundwater*, *54*, 231–242.

Hoekstra, A. Y., & Chapagain, A. K. (2008). *Globalization of water: Sharing the planet's freshwater resources*. Oxford, UK: Blackwell Publishing.

Hoekstra, A. Y., & Mekonnen, M. M. (2012). The water footprint of humanity. *Proceedings of the National Academy of Sciences*, *109*(9), 3232–3237.

Juntakut, P., Snow, D. D., Haacker, E. M. K., & Ray, C. (submitted, January 2018). The long term effect of agricultural, vadose zone and climatic factors on nitrate contamination in Nebraska's groundwater system. Submitted to *Journal of Contaminant Hydrology*.

Martson, L., Konar, M., Cai, X., & Troy, T. J. (2015). Virtual groundwater transfers from overexploited aquifers in the United States. *Proceedings National Academic Sciences USA*, *112*(28), 8561–8566.

McGuire, V. L. (2014). Water-level changes and change in water in storage in the high plains aquifer, predevelopment to 2013 and 2011–13, *U.S. Geological Survey Scientific Investigations Report 2014-5218*, 14 p.

Mekonnen, M. M., & Hoekstra, A. Y. (2012). A global assessment of the water footprint of farm animal products. *Ecosystems*, *15*(3), 401–415.

Nestle, M. (2015). *Soda politics: Taking on big soda (and winning)* (pp. 525). New York: Oxford University Press.

Nolan, J., & Weber, K. A. (2015). Natural uranium contamination in major U.S. *Aquifers Linked to Nitrate, Environmental Science & Technology Letters*, *2*(8), 215–220.

Peck, J. C. (2007). Groundwater management in the high plains aquifer in the USA: Legal problems and innovations, Chapter 14. In M. Giardano & K. Vilholth (Eds.), *Agricultural groundwater revolution: Opportunities and threats to development* (pp. 296–319). Wallingford, UK: CABI.

Polzkill, S., Stejskal, A., Wilke, H., Wistrom, A., Gibson, K., Spiehs, M., & Brozovic, N. (2017). *Paid to pump: How a tax credit could discourage conservation of the high plains aquifer*. Policy Brief #1. Lincoln, NE: Daugherty Water for Food Global Institute. http://waterforfood.nebraska.edu/wp-content/uploads/2017/12/Policy_brief_1_Paid_to_Pump.pdf

Scanlon, B. R., Faunt, C. C., Longuevergne, L., Reedy, R. C., Alley, W. M., McGuire, V. L., & McMahon, P. B. (2012). Groundwater depletion and sustainability of irrigation in the US high plains and central valley. *Proceedings of the National Academy of Sciences*, *109*(24), 9320–9325.

Tilman, D., & Clark, M. (2014). Global diets link environmental sustainability and human health. *Nature, 515*(7528), 518–522.
WWAP (World Water Assessment Programme). (2012). *The United Nations World Water Development Report 4: Managing Water under Uncertainty and Risk.* Paris, UNESCO.

The water footprint of the EU: quantification, sustainability and relevance

Davy Vanham

Introduction

This presentation was delivered on behalf of the European Commission's Joint Research Centre (JRC) at the conference Virtual Water in Agricultural Products: Quantification, Limitations and Trade Policy at the University of Nebraska in Lincoln, Nebraska, USA, 14–16 September 2016, sponsored by the Organisation for Economic Co-operation and Development's (OECD) Co-operative Research Programme. It provides an overview of the quantification, sustainability and relevance of the water footprint (WF) concept, for the case study of the European Union (EU), with a focus on agricultural products. The WF concept provides the opportunity to link the use of water resources to the consumption of goods, thereby addressing supply chain thinking and showing options to remain within planetary boundaries from the perspective of consumers and companies. It provides a strong communication tool for policy-makers and citizens.

The presentation is organized by means of the following sections:

- Water footprint (WF) assessment.
- Sustainability assessment.
- One indicator to be used with others.
- Relevant scenarios.
- Conclusions.

Water footprint (WF) assessment

Introduction

Two different approaches to conduct a WF assessment exist in parallel (see Figure S1 in the supplemental data online). The volumetric approach as described in *The Water Footprint Assessment Manual* (Hoekstra, Chapagain, Aldaya, & Mekonnen, 2011) was first developed and published in 2011. The Life Cycle Assessment (LCA) community

---

The opinions expressed and arguments employed in this publication are the sole responsibility of the authors and do not necessarily reflect those of the OECD or of the governments of its Member countries.
This article has been republished with minor changes. These changes do not impact the academic content of the article.
Supplemental material can be accessed at https://doi.org/10.1080/02508060.2018.1516097.

started later with the development of a methodology in parallel, and published in 2014 its procedure in the ISO 14,046 document (ISO, 2014). The different stages of both methodologies show that both consist of an inventory and impact assessment stage.

The Water Footprint Assessment Manual was developed within the framework of integrated water resources management (IWRM), the ISO 14,046 was not. The latter provides an assessment and reporting standard related to the WF of products, processes and organizations based on LCA. It focuses on product LCAs and environmental impact, while the Water Footprint Network (WFN) standard offers a broader framework in which WFs can be studied with a different focus (product, producer, consumer or geographical) and from different perspectives (environmental sustainability, social equity, resource efficiency or water risk) (Hoekstra 2017). As the present special issue is situated within the context of IWRM, only the WFN approach is discussed here in detail.

Quantification of WF volumes for agriculture

For the quantification of the WF of the EU (the 'inventory' or 'water accounting' phase shown in Figure S1 in the supplemental data online), which refers to a geographical WF (the EU), the following definitions are important:

- WF of production (WF_{prod}): the sum of the direct and indirect water use of domestic freshwater resources of a geographical region.
- WF of consumption (WF_{cons}): the total volume of freshwater used to produce the goods consumed by inhabitants of a geographical region. It is the sum of direct and indirect water use of domestic and foreign water resources through domestic consumption. WF_{cons} equals WF_{prod} plus virtual water (VW) imports (VW_i) but minus virtual water exports (VW_e).
- Consumptive water use: WF amounts relate to consumptive water use (the difference between abstraction/withdrawal and return flow). In addition, these amounts also include the water incorporated into a product along the supply chain.
- Blue water: the liquid water in rivers, lakes, aquifers and reservoirs.
- Green water: the soil water held in the unsaturated zone, formed by precipitation and available to plants (Rockström et al., 2009). Irrigated agriculture receives blue water (from irrigation) as well as green water (from precipitation), while rainfed agriculture receives only green water. The inclusion of green water in IWRM is a necessity and now recommended by most authors (Gerten et al., 2013; Hoekstra, 2016; Jalava, Kummu, Porkka, Siebert, & Varis, 2014; Karimi, Bastiaanssen, & Molden, 2013; Porkka, Gerten, Schaphoff, Siebert, & Kummu, 2016; Ran, Lannerstad, Herrero, Van Middelaar, & De Boer, 2016; Rockström et al., 2014; Schyns, Hoekstra, & Booij, 2015; van Eekelen et al., 2015; Vanham, 2012).
- Grey water: the grey WF component is an indicator of the degree of water pollution (Hoekstra et al., 2011). It is defined as the volume of water needed to dilute a certain amount of pollution such that it meets ambient water quality standards (Hoekstra et al., 2011). The WFN's comprehensive guideline document on accounting the grey WF (Franke, Boyacioglu, & Hoekstra, 2013) states that it is

determined by the pollutant that is most critical (i.e., requires the most water for dilution).

Many authors (e.g., Hoff et al., 2014; Thaler, Zessner, Bertran De Lis, Kreuzinger, & Fehringer, 2012; Vanham & Bidoglio, 2013) regard the grey WF component critically for various reasons:

- The water quantity it represents is not associated with the actual physical water volume of return flows. Therefore, it represents an amount of water that physically cannot be compared/added to blue and green WF components. The latter are actual physical volumes (water flows) within a hydrological water cycle/balance; the grey WF component is not.
- The water quantity is very dependent on data availability and the chosen (available) water quality standard (Thaler et al., 2012). In the past, the grey WF was generally computed based on nitrogen leaching (Hoekstra & Mekonnen, 2012; Mekonnen & Hoekstra, 2015), thereby discarding other water pollutants. Recently, efforts have been made to include other elements, for example, phosphorous (Mekonnen, Lutter, & Martinez, 2016b; Senta, Terzic, & Ahel, 2013) and other chemicals. A recent assessment of the WF of Austria (Thaler et al., 2013) results, for example, in much higher grey WF amounts as compared with a previous assessment (Vanham, 2013b), as the authors included phosphorus in addition to nitrogen. Data availability on pollutants to be included (Franke et al., 2013) is thus a restricting factor.
- When the WF is used in combination with other footprint indicators, the so-called footprint family (Galli et al., 2012; Gephart et al., 2016; Leach et al., 2016), the grey WF could be regarded as double counting. As an example, Vanham, Bouraoui, Leip, Grizzetti, and Bidoglio (2015), who quantified the water and nitrogen footprint of the EU's consumer food waste, chose not to include the grey WF because the nitrogen (N) footprint accounts for N pollution in receiving water bodies.

Although water quality as an issue is very important, we choose not to use the grey WF component, especially in an assessment combined with other footprints. Other scholars include this component in WF assessments.

A quantification of the WF_{prod}, WF_{cons} and VW flows for the EU was presented by Vanham and Bidoglio (2013). Figure S2 (in the supplemental data online) shows that the WF_{prod} for agricultural products adds up to 487 km^3/yr, of which crops constitute 426 km^3/yr (369 km^3/yr green and km^3/yr blue water) and livestock 61 km^3/yr (of which 55 km^3/yr green and 6 km^3/yr blue water). Green water consumption for crop production is spread all over the EU, whereas irrigated blue water consumption is concentrated around the Mediterranean.

The highest livestock water consumption is concentrated in Western Europe.

Figure S3 (in the supplemental data online) shows a VW balance for agricultural products for the EU consisting of WF_{prod}, WF_{cons}, VW_i and VW_e. The WF_{cons} is larger than the WF_{prod}, which means that the EU is a net VW importer for agricultural products. This has several reasons:

- The EU is for some products not self-sufficient (although it is for many products).
- Overconsumption in the EU of water-intensive products (e.g., meat, more than 50% of cereals production in the EU is for feed) (Vanham & Bidoglio, 2013; Vanham, Mekonnen, & Hoekstra, 2013b).
- The production of agricultural products is very water efficient in the EU as compared with the countries from which it imports. Domestic water productivity depends on production methods (irrigated versus rainfed, conservation agriculture, nutrient application etc.), higher yield, and agroclimate conditions (soil, climate etc.) (Vanham & Bidoglio, 2013).

The main agricultural products that account for the largest VW imports include wheat and soybeans from the Americas, cotton from Asia and cocoa and coffee (see Figure S4 in the supplemental data online).

The main agricultural products that account for the largest VW exports from the EU include cereals to Northern Africa, Turkey, Switzerland, China and Japan, as well as meat and wine to the United States, Russia and Japan (see Figure S5 in the supplemental data online).

The status of a region as a net VW importer or exporter says nothing about the sustainability of water use within that region or the regions it imports VW from and/or exports VW to. For that, a WF sustainability assessment needs to be made.

Other important components/sectors to be included in a geographical WF assessment

Although the focus of this paper is on the WF of agriculture, it should be noted that in geographical WF assessments other sectors also account for a WF.

Domestic and industrial water uses are generally included in geographical WF assessments. However, within the water–energy–food–ecosystem (WEFE) nexus context (see Figure S6 in the supplemental data online), other sectors/components should also be included (see Figure S7 online). This is discussed in detail by Vanham (2016). The inclusion of these other components is not common praxis. This is recognized in recent research (e.g., Schyns, Booij, & Hoekstra, 2017). Other scholars refer to the water–energy–food (WEF) nexus (Taniguchi, Endo, Gurdak, & Swarzenski, 2017) or the food–energy–water (FEW) nexus (Chini, Konar, & Stillwell, 2017; Scanlon et al., 2017).

Sustainability assessment

The SDGs for water, energy and food security

In order to provide the Sustainable Development Goals (SDGs) for water, food and energy security in a nexus setting (Vanham, 2016) to a growing and urbanizing global population (UN, 2014), within global planetary boundaries with limited (water) resources availability (Rockstrom et al., 2009), solutions need to come from both the production and consumption sides (Foley et al., 2011;Godfray et al., 2010).

Production-side solutions include (Foley et al., 2011; Godfray et al., 2010):

- The sustainable intensification of agriculture. Increasing water productivity (expressed in, e.g., kg/l, kg/m^3, ton/m^3,, the inverse of WF_{prod}) by closing yield gaps and integrated land and water management on existing agricultural lands (rainfed and irrigated) is key to this development. Often,the increase in water productivity (decrease in WF_{prod}) is referred to as 'more crop per drop'. In the wider context of the WEF nexus (Vanham, 2016), the expression 'more biomass per drop' (apart from crops also fish, livestock, fibre, tree biomass etc.) is more appropriate.
- Production processes along the supply chain need to become more resource efficient and sustainable. Water scarcity and water pollution need to be eradicated.

In order to achieve sustainable WFs in agriculture, water scarcity and the increase in water productivity (or the decrease of WF_{prod}) thus need to be addressed. Both should be an integral part of a WF sustainability assessment. It is also referred to as SDG target 6.4: 'By 2030, substantially increase water-use efficiency across all sectors and ensure sustainable withdrawals and supply of freshwater to address water scarcity and substantially reduce the number of people suffering from water scarcity.'

Demand-side solutions include:

- Consumer behaviour in water use, energy consumption, food consumption and (food) waste generation.
- Some scholars also state that family planning to lower per capita fertility should be part of the solution (Speidel, Weiss, Ethelston, & Gilbert, 2009). In many regions of the world, significant population increases are projected for the next few decades (UN, 2014). There are consequently more frequent calls to address environmental problems by advocating further reductions in human fertility (Bradshaw & Brook, 2014), including in the second notice of 'World Scientists' Warning to Humanity' (Ripple et al., 2017).

It is clear that the provision of water, energy and food security for people within a geographical region relates to the consumption side of this geographical region and not the production side. Trade between regions is thus essential to provide these three securities. Sustainability does not mean self-sufficiency.

Water-related sustainability of EU agricultural production

Many regions of the EU are confronted with physical (blue) water scarcity and/or groundwater depletion at least for some part of the year (Custodio et al., 2016; Hoekstra, Mekonnen, Chapagain, Mathews, & Richter, 2012; Mekonnen & Hoekstra, 2016; Wada, van Beek, & Bierkens, 2012; Werner et al., 2013). In addition, the EU imports agricultural products from regions where physical (blue) water scarcity and/or groundwater depletion occurs (Dalin, Wada, Kastner, & Puma, 2017; Mekonnen & Hoekstra, 2016; Wada et al., 2012).

Within the EU, agricultural production of crops and livestock attributes to pollution (e.g., nutrients) of freshwater bodies (see Figure S8 in the supplemental data online) and bordering seas (Bouraoui and Grizzetti, 2011; Bouraoui, Thieu,

Grizzetti, Britz, & Bidoglio, 2014; Grizzetti, Bouraoui, & Aloe, 2012; Leip, Weiss, Lesschen, & Westhoek, 2014) (see Figure S9 online). In addition, agricultural products imported into the EU contribute to the pollution of freshwater bodies in regions where they are produced (Mekonnen & Hoekstra, 2015).

Eastern Europe is identified as a region with considerable intensification opportunities for major cereals (Mueller et al., 2012). Amongst the greatest opportunities for increases in global absolute production (from closing yield gaps to 100% of estimated attainable yields) is wheat in Eastern Europe (predominately due to nutrient limitation). Romania, for example, has average wheat yields of 2–3 ton/ha, whereas in Ireland, average yields are in the range 7–8 ton/ha (see Figure S10 in the supplemental data online). The WF_{prod} of many crops can therefore be decreased (water productivity increased) by integrated land and water management (including nutrient management), as the example of wheat shows (see Figure S10 online). Ireland, for example, has amongst the lowest wheat WF_{prod} in the world (460 m^3/ton), whereas in Romania it averages 1980 m^3/ton.

A WF sustainability assessment is an integral part of a WF assessment (see Figure S1 in the supplemental data online). To make a WF sustainability assessment, theoretically environmental, social and economic indicators are required. This is discussed in more detail by Hoekstra et al. (2011) and Vanham and Bidoglio (2013). The indicators used for an environmental impact assessment are discussed below. The increase of agricultural water productivity as part of a WF sustainability assessment, is briefly discussed below.

To increase water productivity for agricultural production and tackle water scarcity and water pollution, integrated land and water management is a necessity. This refers to the promotion of system-based farming practices that integrate land, water, nutrient, livestock and crop management within a river basin management approach. The following farm-level actions are relevant (amongst others) (Vanham & Bidoglio, 2013):

- Management of irrigation systems: choice of technology (e.g., drip), irrigation scheduling, deficit irrigation, use of soil moisture and canopy sensors.
- Soil moisture conservation: e.g., by conservation tillage that optimizes soil carbon content or mulching (organic residues, plastic films).
- Nutrient management: including on-farm nutrient cycling or precision agriculture, to increase yields and water productivity and reduce water pollution.
- Cropping strategies: including the avoidance of growing high-water-demand crops during summer (e.g., maize) when water availability is low; and growing drought-tolerant or low-water-demand crops.
- Optimal use of available water resources: including rain water harvesting, wastewater recycling and artificial groundwater recharge (natural water buffer for use during dry periods).
- Limit the intensification of livestock production to animal welfare constraints.
- Riparian buffers (Weissteiner, Bouraoui, & Aloe, 2013) or constructed wastewater treatment wetlands to control water quality.

Environmental impact assessment of WF_{prod}

For an environmental impact assessment, the following indicators are required:

- Blue water scarcity: generally defined as the ratio between water use (water abstraction or consumption) and water availability (total or ecological, the latter being total water availability minus ecological/environmental flows). It has been used in different variations (Alcamo, Flörke, & Märker, 2007; EEA, 2003; Mekonnen & Hoekstra, 2016; Raskin, Gleick, Kirshen, Pontius, & Strzepek, 1997; Rijsberman, 2006; Smakhtin, Revenga, & Döll, 2004; Vanham, Fleischhacker, & Rauch, 2009a,2009b; Wada et al., 2011). It is now generally recognized that also environmental flows need to be taken into account (Vanham & Bidoglio, 2013). In addition, the sustainability of groundwater use needs to be addressed. Groundwater is often connected to surface water. In such connected systems, river surface water can feed groundwater, and vice versa. When groundwater levels sink below riverbeds, the connection is generally lost. This is often the case in heavily depleted aquifers. Specific indicators addressing the sustainability of groundwater use have been developed (e.g., Gleeson, Wada, Bierkens, & van Beek, 2012; Wada & Bierkens, 2014). Indicator 6.4.2 of the SDGs ('Level of water stress') is a recommended metric to assess blue water scarcity. However, with respect to the WF, water consumption and not water abstraction needs to be used (Vanham et al., 2018a).
- Green water scarcity/impact of green WF: there is no clear recommendation how this should be addressed. An overview of possible relevant indicators/assessments was recently made by (Schyns et al., 2015). Some authors argue that a green WF impact assessment refers to the difference in green ET between current agricultural land use (crop or livestock production) and a reference state.
- Water pollution level: as we do not use grey water, this also applies to this indicator.

Water productivity

A setting to achieve higher water productivity for crops is achieved by determining benchmarks. A global assessment of WF benchmarks for crops was made by Mekonnen and Hoekstra (2014). Crop WF benchmarks are, however, dependent on agro-climatological conditions.

The terminology 'more biomass per drop' refers to other sectors than only crop production; it also includes livestock production, aquaculture or tree biomass (apart from crops also fish, livestock, fibre, tree biomass etc.).

A major trade-off in increasing water productivity, e.g., in nutrient-limited areas, is a potential increase in water pollution. When comparing Figure S8 (and Figure S2b as an indicator for the location of livestock production in the EU) with Figure S10b (all in the supplemental data online), it is clear that high water productivity in reality often results in high water pollution.

One indicator to be used with others

The indicator WF provides information on the use of the resource water along the supply chain as well as the impact of this water use. However, it is a partial tool. It does not account for water aspects such as flooding or lack of water infrastructure, nor does it account for other (finite) resources and the (environmental) impact of their use. An overview of resources required to provide the three essential human securities and associated environmental impacts is displayed in Figure S11 in the supplemental data online.

For integrated policy options, more indicators are required. Different authors therefore recommend using a family of environmental footprint indicators (Galli et al., 2012; Ibidhi, Hoekstra, Gerbens-Leenes, & Chouchane, 2017), the latter being an umbrella term for the different footprint concepts that have been developed during the last two decades (Hoekstra & Wiedmann, 2014). Other well-established footprint indicators include the nitrogen footprint (Leip et al., 2014), phosphorous footprint (van Dijk, Lesschen, & Oenema, 2016), land footprint (Weinzettel, Hertwich, Peters, Steen-Olsen, & Galli, 2013), carbon footprint and energy footprint. Indeed, policy recommendations developed based on one footprint indicator can make no sense when an additional indicator is analyzed (Mekonnen, Gerbens-Leenes, & Hoekstra, 2016a). In parallel, the LCA community developed methodologies that also include different resources and impacts, including for the resource water. To summarize:

- It is not recommended to develop integrated policy options based on the WF alone. Footprint families are better suited, possibly in addition to other indicators.
- The nitrogen and phosphorous footprints include a quantification of the impact on water pollution by use of these nutrients. As we do not use the grey WF component, these footprints account for this important pollution source.
- It is not recommended to develop trade policies based on only VW trade analyses.
- Risks in trade flows can, however, be detected, for example, for importing countries that import agricultural goods (partially) produced unsustainably (Hoekstra & Mekonnen, 2016; Vanham, Gawlik, & Bidoglio, 2017a) or with depleted groundwater (Dalin et al., 2017).
- There is a need to streamline the footprint and LCA methodologies.
- The strength of the LCA is that it integrates different resources and impacts in its analysis framework.

Relevant scenarios

Apart from scenario analysis on the production side (e.g., crop WF_{prod} benchmarking, as previously described), the WF provides a unique possibility to link the resource water with the consumption side, and thus scenarios on the consumption side. During the last years, the JRC has made several such analyses:

- Lost water resources due to EU consumer food waste (Vanham et al., 2015).
- The effect of different diets on the EU's WF_{cons} (Vanham et al., 2013b) (see Figure S12 in the supplemental data online).
- The effect of diets on the WF of different EU zones (Vanham, Hoekstra, & Bidoglio, 2013a), countries (Vanham, 2013b), EU river basins (Vanham, 2013a; Vanham & Bidoglio, 2014a) and cities (Mekonnen et al., 2016a; Vanham & Bidoglio, 2014b; Vanham, Gawlik, & Bidoglio, 2017b; Vanham et al., 2016a; Vanham, Mak, & Gawlik, 2016b; Vanham, Comero, Gawlik, & Bidoglio, 2018b).

Figure S12 online shows the following:

- EU citizens can save a lot of water by shifting to a healthier diet. Currently they consume too much sugar, oil, meat, animal fat and alcoholic beverages and not enough fruit and vegetables.
- The WF of food consumption is of a whole other magnitude than the amounts in domestic water use (130 l per capita per day EU average).
- In the past, information campaigns focused on the reduction of direct water use, with resulting effects, as displayed for the Netherlands. Also, the industry made important efforts to make, for example, washing machines or toilets, more water efficient.
- Water is probably the most recognizable resource for consumers. Consumers really understand water amounts (e.g., litres) because they are confronted with them in daily life, as opposed to numbers/amounts of other footprints/resources such as the carbon or nitrogen footprints. This is part of the success of the WF.
- The WF is a very strong communication tool and is also valuable for awareness raising.

As overweight and obesity have been steadily increasing over the last decades to currently alarming rates in Organisation for Economic Co-operation and Development (OECD) countries (see Figure S13 in the supplemental data online) (NCD-RisC, 2016), a shift to healthier diets would mean the following:

- A massive positive impact on the overall health of the population.
- Reduce the WF of OECD countries, because the agricultural products that are over-consumed, such as meat and sugar, are very water intensive to produce.
- Have the potential to reduce health spending in OECD countries.

Conclusions

The EU is a region where both water scarcity and pollution due to agricultural processes prevail. Both are, however, spatially and temporally very heterogeneously spread. The western zone is, for example, characterized by the highest wheat water productivities in the world, whereas the eastern zone is identified as a hotspot for closing wheat yield gaps and increasing water productivity. In addition, the EU imports agricultural products from regions outside its borders affected by water scarcity and pollution.

With respect to the title of the conference 'Virtual Water in Agricultural Products: Quantification, Limitations and Trade Policy', it can be stated that trade is essential to provide for the SDGs' food, energy and water security in a sustainable way, within a nexus (the water–energy–food, or WEF nexus) context (Vanham, 2016). Water is one resource to provide for these securities, but not the only one. Water scarcity and water pollution are two important impacts as a result of providing these securities, but not the only ones (see Figure S11 in the supplemental data online). Therefore, it can be concluded that for integrated policy recommendations, including trade policies, more indicators than only the WF need to be used.

There are two main approaches in WF assessment being developed, one described in the *Water Footprint Manual* (Hoekstra et al., 2011), which is developed within the framework of the IWRM, the other described in ISO 14,046 (ISO, 2014), which is developed by the LCA community. Both are based on supply-chain or life-cycle thinking, accounting for all water uses along the supply chain. Both include an inventory and impact assessment phase (see Figure S1 in the supplemental data online). Synergies between both include the following:

- A strength of the LCA approach is that it takes into account different resources and impacts, not only for water. For integrated policy recommendations, such an integrated approach is necessary. For the latter goal, the *Water Footprint Manual* (Hoekstra et al., 2011) approach should be used in combination with other indicators. Footprint family assessments have been done in the past for particular topics and are recommended. In addition, this would partially compensate for the grey WF, as, for example, the nitrogen and phosphorous footprints would partially compensate for the impacts on water quality.
- The inclusion of green water as a WF component is a necessity. This has been from the beginning included in the *Water Footprint Manual*, but has not originally been advocated by the LCA approach.
- For sustainability assessments, also a water productivity assessment of agricultural production is a necessity due to global challenges and SDGs. As increases in water productivity (decreases in WF_{prod}) can lead to increased water pollution (trade-offs), a simultaneous assessment of other indicators is necessary. This is also expressed in SDG target 6.4.
- The indicator blue water scarcity has been developed and used within the water management community for a few decades. As a method, the new SDG indicator 6.4.2 is recommended, but – in the case of a WF assessment – by using water consumption instead of water withdrawal (Vanham et al., 2018a).

For geographical WF assessments, apart from the WF of agricultural products, also other WF components from other sectors need to accounted for, as described by Vanham (2016).

The WF concept provides a unique opportunity to link the use of water resources to the consumption of goods. In order to achieve the three main human securities in a sustainable way, solutions also need to come from the consumption side, especially – regarding food – through the diet and food waste behaviour of global consumers. Therefore, the main strengths of the WF concept are the following:

- A visualization of the link between (local) consumption and (global) appropriation.
- As water is probably the most recognizable of our resources in terms of quantities (the average consumer has a very good idea of, for example, the volume of 1 litre of water), the WF is very valuable for communication and awareness raising of water appropriation related to consumed goods.
- Especially in OECD countries, where overweight and obesity with associated health problems have been increasing steadily during the last decades, a shift to

a healthier diet of the general population would be of great benefit for general health, healthcare spending but also resource use and the environment. The WF provides for an additional indicator to show this necessity to policy-makers and the general public.

For the latter reasons, the European Commission has started specific communications using the WF indicator. It is, for example, part of the recently published *Urban Water Atlas of Europe* (European Commission, 2017; Gawlik et al., 2017), which was presented on 27 April 2017 during the meeting of ministers in charge of water management from the 43 members of the Union for the Mediterranean (UfM), hosted by the Maltese government in Valetta (UfM, 2017).

Acknowledgements

This paper was given at the workshop on Virtual Water in Agricultural Products: Quantification, Limitations and Trade Policy (Lincoln, Nebraska, USA, 14-16 September 2016), which was sponsored by the OECD Co-operative Research Programme: Biological Resource Management for Sustainable Agricultural Systems whose financial support made it possible for the author to participate in the workshop.

Disclosure

No potential conflict of interest was reported by the authors.

ORCID

Davy Vanham http://orcid.org/0000-0002-7294-7979

References

Alcamo, J., Flörke, M., & Märker, M. (2007). Future long-term changes in global water resources driven by socio-economic and climatic changes. *Hydrological Sciences Journal*, *52*(2), 247–275.

Bouraoui, F., & Grizzetti, B. (2011). Long term change of nutrient concentrations of rivers discharging in European seas. *Science of the Total Environment*, *409*(23), 4899–4916.

Bouraoui, F., Thieu, V., Grizzetti, B., Britz, W., & Bidoglio, G. (2014). Scenario analysis for nutrient emission reduction in the European inland waters. *Environmental Research Letters*, *9* (12), 125007.

Bradshaw, C. J. A., & Brook, B. W. (2014). Human population reduction is not a quick fix for environmental problems. *Proceedings of the National Academy of Sciences*, *111*(46), 16610–16615.

Chini, C. M., Konar, M., & Stillwell, A. S. (2017). Direct and indirect urban water footprints of the United States. *Water Resources Research*, *53*(1), 316–327.

Custodio, E., Andreu-Rodes, J. M., Aragón, R., Estrela, T., Ferrer, J., García-Aróstegui, J. L., … & del Villar, A. (2016). Groundwater intensive use and mining in south-eastern peninsular Spain: Hydrogeological, economic and social aspects. *Science of the Total Environment*, *559*, 302–316.

Dalin, C., Wada, Y., Kastner, T., & Puma, M. J. (2017). Groundwater depletion embedded in international food trade. *Nature*, *543*(7647), 700–704.

EEA, 2003. Water exploitation index, Author, Copenhagen. Retrieved from: http://www.eea.europa.eu/data-and-maps/indicators/waterexploitation-index.

European Commission, 2017. Press release: Urban Water Atlas for Europe – a 360° view of water management in cities, Brussels: Retrieved from http://europa.eu/rapid/press-release_IP-17-1110_en.htm.

Foley, J. A., Ramankutty, N., Brauman, K. A., Cassidy, E. S., Gerber, J. S., Johnston, M., … & Balzer, C (2011). Solutions for a cultivated planet. *Nature, 478*(7369), 337–342.

Franke, N. A., Boyacioglu, H., & Hoekstra, A. Y. (2013). *Grey water footprint accounting: Tier 1 supporting guidelines, UNESCO-IHE.* Delft: Institute for Water Education.

Galli, A., Wiedmann, T., Ercin, E., Knoblauch, D., Ewing, B., & Giljum, S (2012). Integrating ecological, carbon and water footprint into a 'Footprint Family' of indicators. Definition and role in tracking human pressure on the planet. *Ecological Indicators, 16*, 100–112.

Gawlik, B. M., Easton, P., Koop, S., van Leeuwen, K., & Elelman, R. (2017). *Urban water atlas for Europe.* Luxemburg: European Commission, Publication Office of the European Union.

Gephart, J. A., Davis, K. F., Emery, K. A., Leach, A. M., Galloway, J. N., & Pace, M. L. (2016). The environmental cost of subsistence: Optimizing diets to minimize footprints. *Science of the Total Environment, 553*, 120–127.

Gerten, D., Hoff, H., Rockström, J., Jägermeyr, J., Kummu, M., & Pastor, A. V. (2013). Towards a revised planetary boundary for consumptive freshwater use: Role of environmental flow requirements. *Current Opinion in Environmental Sustainability, 5*(6), 551–558.

Gleeson, T., Wada, Y., Bierkens, M. F. P., & van Beek, L. P. H. (2012). Water balance of global aquifers revealed by groundwater footprint. *Nature, 488*(7410), 197–200.

Godfray, H. C. J., Beddington, J. R., Crute, I. R., Haddad, L., Lawrence, D., Muir, J. F., … & Toulmin, C. (2010). Food security: The challenge of feeding 9 billion people. *Science, 327*, 812–818.

Grizzetti, B., Bouraoui, F., & Aloe, A. (2012). Changes of nitrogen and phosphorus loads to European seas. *Global Change Biology, 18*(2), 769–782.

Hoekstra, A. Y. (2016). A critique on the water-scarcity weighted water footprint in LCA. *Ecological Indicators, 66*, 564–573.

Hoekstra, A.Y. (2017). Water footprint assessment: Evolvement of a new research field. *Water Resources Management, 31*(10), 3061–81.

Hoekstra, A. Y., Chapagain, A. K., Aldaya, M. M., & Mekonnen, M. M. (2011). *The water footprint assessment manual: Setting the global standard.* London, UK: Earthscan.

Hoekstra, A. Y., & Mekonnen, M. M. (2012). The water footprint of humanity. *Proceedings National Academic Sciences USA, 109*, 3232–3237.

Hoekstra, A. Y., & Mekonnen, M. M. (2016). Imported water risk: The case of the UK. *Environmental Research Letters, 11*(5), 055002.

Hoekstra, A. Y., Mekonnen, M. M., Chapagain, A. K., Mathews, R. E., & Richter, B. D. (2012). Global monthly water scarcity: Blue water footprints versus blue water availability. *PLoS ONE, 7*(2), e32688.

Hoekstra, A. Y., & Wiedmann, T. O. (2014). Humanity's unsustainable environmental footprint. *Science, 344*(6188), 1114–1117.

Hoff, H., et al. (2014). Water footprints of cities: Indicators for sustainable consumption and production. *Hydrology and Earth System Sciences, 18*(1), 213–226.

Ibidhi, R., Hoekstra, A. Y., Gerbens-Leenes, P. W., & Chouchane, H. (2017). Water, land and carbon footprints of sheep and chicken meat produced in Tunisia under different farming systems. *Ecological Indicators, 77*, 304–313.

ISO. (2014). *ISO 14046 – Environmental management – Water footprint – Principles, requirements and guidelines.* Geneva: International Organization for Standardization.

Jalava, M., Kummu, M., Porkka, M., Siebert, S., & Varis, O. (2014). Diet change—A solution to reduce water use?. *Environmental Researcher Letters, 9*(7), 074016.

Karimi, P., Bastiaanssen, W. G. M., & Molden, D. (2013). Water Accounting Plus (WA+): A water accounting procedure for complex river basins based on satellite measurements. *Hydrology and Earth System Sciences, 17*(7), 2459–2472.

Leach, A. M., Emery, K. A., Gephart, J., Davis, K. F., Erisman, J. W., Leip, A., … & Castner, E. (2016). Environmental impact food labels combining carbon, nitrogen, and water footprints. *Food Policy*, *61*, 213–223.

Leip, A., Weiss, F., Lesschen, J. P., & Westhoek, H. (2014). The nitrogen footprint of food products in the European Union. *The Journal of Agricultural Science*, *152*(S1), 20–33.

Mekonnen, M. M., Gerbens-Leenes, P. W., & Hoekstra, A. Y. (2016a). Future electricity: The challenge of reducing both carbon and water footprint. *Science of the Total Environment*, *569–570*, 1282–1288.

Mekonnen, M. M., & Hoekstra, A. Y. (2011). The green, blue and grey water footprint of crops and derived crop products. *Hydrology and Earth System Sciences*, *15*, 1577–1600.

Mekonnen, M. M., & Hoekstra, A. Y. (2014). Water footprint benchmarks for crop production: A first global assessment. *Ecological Indicators*, *46*, 214–223.

Mekonnen, M. M., & Hoekstra, A. Y. (2015). Global gray water footprint and water pollution levels related to anthropogenic nitrogen loads to fresh water. *Environmental Science & Technology*, *49*(21), 12860–12868.

Mekonnen, M. M., & Hoekstra, A. Y. (2016). Four billion people facing severe water scarcity. *Science Advances*, *2*, 2.

Mekonnen, M. M., Lutter, S., & Martinez, A. (2016b). Anthropogenic nitrogen and phosphorus emissions and related grey water footprints caused by EU-27's crop production and consumption. *Water*, *8*, 30.

Monfreda, C., Ramankutty, N., & Foley, J. A. (2008). Farming the planet: 2. Geographic distribution of crop areas, yields, physiological types, and net primary production in the year 2000. *Global Biogeochemical Cycles*, *22*(1).

Mueller, N. D., Gerber, J. S., Johnston, M., Ray, D. K., Ramankutty, N., & Foley, J. A. (2012). Closing yield gaps through nutrient and water management. *Nature*, *490*(7419), 254–257.

NCD-RisC. (2016). Trends in adult body-mass index in 200 countries from 1975 to 2014: A pooled analysis of 1698 population-based measurement studies with 19.2 million participants. *The Lancet*, *387*(10026), 1377–1396.

Porkka, M., Gerten, D., Schaphoff, S., Siebert, S., & Kummu, M. (2016). Causes and trends of water scarcity in food production. *Environmental Research Letters*, *11*(1), 015001.

Ran, Y., Lannerstad, M., Herrero, M., Van Middelaar, C. E., & De Boer, I. J. M. (2016). Assessing water resource use in livestock production: A review of methods. *Livestock Science*, *187*, 68–79.

Raskin, P., Gleick, P., Kirshen, P., Pontius, G., & Strzepek, K. (1997). *Water futures: Assessment of long-range patterns and problems. Comprehensive assessment of the freshwater resources of the world. SEI REPORT*. Stockholm, Sweden: Stockholm Environment Institute.

Rijsberman, F. R. (2006). Water scarcity: Fact or fiction?. *Agricultural Water Management*, *80*(1–3), 5–22.

Ripple, W. J., Wolf, C., Newsome, T. M., Galetti, M., Alamgir, M., Crist, E., ... & Laurance, W.F. (2017). World scientists' warning to humanity: a second notice. *Bioscience*, *67*(12), 1026–28.

Rockström, J.,Falkenmark, M., Allan, T., Folke, C., Gordon, L., Jägerskog, A., … & Postel, S. (2014). The unfolding water drama in the Anthropocene: Towards a resilience-based perspective on water for global sustainability. *Ecohydrology*, *7*(5), 1249–1261.

Rockström, J., Falkenmark, M., Karlberg, L., Hoff, H., Rost, S., & Gerten, D. (2009). Future water availability for global food production: The potential of green water for increasing resilience to global change. *Water Resources Research*, *45*, W00A12.

Rockstrom, J., Steffen, W., Noone, K., Persson, Å., Chapin F. S. III, Lambin, E. F., … & Nykvist, B (2009). A safe operating space for humanity. *Nature*, *461*(7263), 472–475. .

Scanlon, B. R., Ruddell, B. L., Reed, P. M., Hook, R. I., Zheng, C., Tidwell, V. C., & Siebert, S (2017). The food–energy–water nexus: Transforming science for society. *Water Resources Research*, *53*(5), 3550–3556.

Schyns, J. F., Booij, M. J., & Hoekstra, A. Y. (2017). The water footprint of wood for lumber, pulp, paper, fuel and firewood. *Advances in Water Resources*. doi:10.1016/j.advwatres.2017.05.013

Schyns, J. F., Hoekstra, A. Y., & Booij, M. J. (2015). Review and classification of indicators of green water availability and scarcity. Hydrological Earth System *Science*, *19*(11), 4581–4608.

Senta, I., Terzic, S., & Ahel, M. (2013). Occurrence and fate of dissolved and particulate antimicrobials in municipal wastewater treatment. *Water Research*, *47*(2), 705–714.

Smakhtin, V., Revenga, C., & Döll, P. (2004). *Taking into account environmental water requirements in global-scale water resources assessments*. Comprehensive Assessment Research Report 2. Colombo, Sri Lanka: Comprehensive Assessment Secretariat.

Speidel, J. J., Weiss, D. C., Ethelston, S. A., & Gilbert, S. M. (2009). Population policies, programmes and the environment. *Philosophical Transactions of the Royal Society B: Biological Sciences*, *364*(1532), 3049–3065.

Taniguchi, M., Endo, A., Gurdak, J. J., & Swarzenski, P. (2017). Water–energy–food nexus in the Asia-Pacific region. Journal of Hydrology *Regional Studies*, *11*, 1–8.

Thaler, S., Zessner, M., Bertran De Lis, F., Kreuzinger, N., & Fehringer, R. (2012). Considerations on methodological challenges for water footprint calculations. *Water Science and Technology*, *65*(7), 1258–1264.

Thaler, S., Zessner, M., Mayr, M. M., Haider, T., Kroiss, H., & Rechberger, H. (2013). Impacts of human nutrition on land use, nutrient balances and water consumption in Austria. *Sustainability of Water Quality and Ecology*, *1–2*, 24–39.

UfM, 2017. Press release: UfM Ministers agree on new framework for an enhanced regional cooperation on water in the Mediterranean, Retrieved from: http://ufmsecretariat.org/ufm-ministersagree-on-new-framework-for-an-enhanced-regional-cooperation-on-water-in-theme diterranean-2/.

UN. (2014). *World urbanization prospects: The 2014 revision, highlights*. New York: United Nations.

van Dijk, K. C., Lesschen, J. P., & Oenema, O. (2016). Phosphorus flows and balances of the European Union member states. *Science of the Total Environment*, *542*, 1078–1093.

van Eekelen, M. W., Bastiaanssen, W. G., Jarmain, C., Jackson, B., Ferreira, F., Van der Zaag, P., ... & Dost, R. J. J. (2015). A novel approach to estimate direct and indirect water withdrawals from satellite measurements: A case study from the Incomati basin. *Agriculture, Ecosystems & Environment*, *200*, 126–142.

Vanham, D., Fleischhacker, E., & Rauch, W. (2009a). Impact of an extreme dry and hot summer on water supply security in an alpine region. *Water Science and Technology*, *59*(3), 469–477.

Vanham, D., Fleischhacker, E., & Rauch, W. (2009b). Impact of snowmaking on alpine water resources management under present and climate change conditions. *Water Science and Technology*, *59*(9), 1793–1801.

Vanham, D. (2012). A holistic water balance of Austria – How does the quantitative proportion of urban water requirements relate to other users? *Water Science and Technology*, *66*(3), 549–555.

Vanham, D., & Bidoglio, G. (2013). A review on the indicator water footprint for the EU28. *Ecological Indicators*, *26*, 61–75.

Vanham, D. (2013a). An assessment of the virtual water balance for agricultural products in EU river basins. *Water Resources and Industry*, *1–2*, 49–59.

Vanham, D. (2013b). The water footprint of Austria for different diets. *Water Science and Technology*, *67*(4), 824–830.

Vanham, D., Hoekstra, A. Y., & Bidoglio, G. (2013a). Potential water saving through changes in European diets. *Environment International*, *61*, 45–56.

Vanham, D., Mekonnen, M. M., & Hoekstra, A. Y. (2013b). The water footprint of the EU for different diets. *Ecological Indicators*, *32*, 1–8.

Vanham, D., & Bidoglio, G. (2014a). The water footprint of agricultural products in European river basins. *Environmental Research Letters*, *9*(6), 064007.

Vanham, D., & Bidoglio, G. (2014b). The water footprint of Milan. *Water Science and Technology*, *69*(4), 789795.

Vanham, D., Bouraoui, F., Leip, A., Grizzetti, B., & Bidoglio, G. (2015). Lost water and nitrogen resources due to EU consumer food waste. *Environmental Research Letters, 10*(8), 084008.
Vanham, D. (2016). Does the water footprint concept provide relevant information to address the water–Food–Energy–Ecosystem nexus? . *Ecosystem Services, 17*, 298–307.
Vanham, D., Del Pozo, S., Pekcan, A. G., Keinan-Boker, L., Trichopoulou, A., & Gawlik, B. M (2016a). Water consumption related to different diets in Mediterranean cities. *Science of the Total Environment, 573*, 96–105.
Vanham, D., Mak, T. N., & Gawlik, B. M. (2016b). Urban food consumption and associated water resources: The example of Dutch cities. *Science of the Total Environment, 565*, 232–239.
Vanham, D., Gawlik, B. M., & Bidoglio, G. (2017a). *Cities as hotspots of indirect water consumption: The case study of Hong Kong.* doi:10.1016/j.jhydrol.2017.12.004.
Vanham, D., Gawlik, B. M., & Bidoglio, G. (2017b). Food consumption and related water resources in Nordic cities. *Ecological Indicators, 74*, 119–129.
Vanham, D., Hoekstra, A. Y., Wada, Y., Bouraoui, F., de Roo, A., Mekonnen, M. M., … & Kummu, M. (2018a). Physical water scarcity metrics for monitoring progress towards sustainable development goal target 6.4: An evaluation of indicator 6.4.2 'Level of water stress'. *Science of the Total Environment*, 613-614, 218-232. doi: 10.1016/j.scitotenv.2017.09.056.
Vanham, D., Comero, S., Gawlik, B. M., & Bidoglio, G. (2018b). The water footprint of different diets within European sub-national geographical entities. *Nature Sustainability, 1*, 518–525. doi: 10.1038/s41893-018-0133-x
Wada, Y., & Bierkens, M. (2014). Sustainability of global water use: Past reconstruction and future projections. *Environmental Research Letters, 9*, 104003.
Wada, Y., van Beek, L. P. H., & Bierkens, M. F. P. (2012). Nonsustainable groundwater sustaining irrigation: A global assessment. *Water Resources Research, 48*(6).
Wada, Y., van Beek, L. P. H., Viviroli, D., Dürr, H. H., Weingartner, R., & Bierkens, M. F. (2011). Global monthly water stress: 2. Water demand and severity of water stress. *Water Resources Research, 47*(7).
Weinzettel, J., Hertwich, E. G., Peters, G. P., Steen-Olsen, K., & Galli, A. (2013). Affluence drives the global displacement of land use. *Global Environmental Change, 23*(2), 433–438.
Weissteiner, C. J., Bouraoui, F., & Aloe, A. (2013). Reduction of nitrogen and phosphorus loads to European rivers by riparian buffer zones. *Knowledge and Management of Aquatic Ecosystems, 408*, 08.
Werner, A. D., et al. (2013). An initial inventory and indexation of groundwater mega-depletion cases. *Water Resources Manage, 27*(2), 507–533.

The exposure of a fresh fruit and vegetable supply chain to global water-related risks

Tim Hess and Chloe Sutcliffe

ABSTRACT

We have combined estimates of the UK's supply of fresh fruit and vegetables (1996 – 2015) with estimates of water requirements and water scarcity in producing countries, to identify where the supply is exposed to physical, regulatory and reputational water risks and how this has changed over time. Some 76% of the freshwater consumed in the supply of fresh fruit and vegetables to the UK is withdrawn overseas. The supply chain is particularly exposed to water risks in Spain, Egypt, South Africa, Chile, Morocco, Israel and Peru. Exposure has increased over time.

Introduction

The production of fresh fruit and vegetables (FF&V) requires large quantities of water. With population growth, dietary change and the greater recognition of the importance of fruit and vegetable consumption for health, demand is increasing, and the areas where FF&V are cultivated have been extending, particularly in semi-arid and arid environments where water shortages are common (Kumar, Rouphael, Cardarelli, & Colla, 2017).

Water is used in many stages of the FF&V value chain including growing (rainfall, irrigation), processing (washing produce, pack-house wash-down, sanitation) and distribution (wash-down). As such, a secure and reliable supply of water of an appropriate quality is critical to businesses that grow and process fruit and vegetables, and consequently to the security of supply to retailers and ultimately consumers.

FF&V value chains are exposed to different types of water risk. In this article we focus predominantly on the physical, regulatory and reputational risks.

Physical risks such as drought are related to physical access to water. Drought may mean that water is not available for irrigation of FF&V, leading to plant water stress that during critical growth periods (which vary by species) can adversely affect a plant's ability to take up nutrients, limiting growth, yield and affecting the quality of the resulting produce (Kumar et al., 2017). At the extreme, water stress may lead to the abortion of flowers and eventually plant death (Ripoll et al., 2014). This can have potentially severe economic consequences for stakeholders within the value chain.

The opinions expressed and arguments employed in this publication are the sole responsibility of the authors and do not necessarily reflect those of the OECD or of the governments of its Member countries.

Supplementary data can be accessed here

Drought in California in 2015 was estimated to have caused 220,000 ha to be fallowed and led to total vegetable crop revenue losses of $93 million (Howitt, Medellín-Azuara, & Macewan, 2014). Physical water risks may also arise due to biological or chemical quality degradation. This problem has been projected to increase in future due to the impacts of climate change (Van Vliet et al., 2013).

Regulatory risk refers primarily to reallocation of supply by water authorities to other sectors or environmental uses. Globally, more than 70% of water withdrawals are for agricultural use (FAO, 2017), and agriculture has been identified as the cause of water depletion in many river basins (Vörösmarty et al., 2010). Consequently, water resource management policies, such as the EU Water Framework Directive (EC, 2000), often encourage the reduction of water use in agriculture in order to release, or secure, water for other sectors (especially the environment) through removal, restriction or modification of water rights. In times of drought, temporary restrictions of water withdrawals for agriculture are usually enforced before physical water scarcity becomes a constraint in order to protect access to water for higher priority sectors. In this way, the FF&V value chain is exposed to a regulatory risk that increases with physical water scarcity.

A *reputational risk* can arise from actual, or perceived, negative impacts of production on local water scarcity. In the UK, articles have appeared in the national press linking withdrawal of water for FF&V to negative impacts on water supply for local communities (e.g., production of Peruvian asparagus, Lawrence, 2010) or environmental degradation (e.g., Spanish strawberries, Fotheringham, 2012). These reports may influence consumer demand or lead to litigation risk which could ultimately remove a business's 'licence to operate' in a particular locality. Such reputational risks are likely to increase as growing over-exploitation of water resources has negative impacts on other water users and the ecology.

By far the greatest volumes of water used in the value chain for FF&V are associated with growing crops (Mekonnen and Hoekstra, 2011). As such, in this article we focus predominantly on these three types of risk in the location of production.

Water for FF&V production may be supplied by rainfall ('green' water) or as water withdrawn from rivers, lakes and groundwater ('blue' water) (Falkenmark, 1995). Rainfed crops are exposed to weather variability and climate change risks; these have been considered elsewhere (e.g., Knox, Daccache, Hess, & Haro, 2016; Knox, Hess, Daccache, & Wheeler, 2012), and are beyond the scope of this study. However, as blue water has alternative and competing uses, in this paper, we focus on the risks associated with the use of blue water in the growing of FF&V.

Blue water is primarily used for irrigation of FF&V, although very small quantities are also used for other production-related on-farm activities, such as crop spraying. The increasing reliance of FF&V on irrigation reflects the global drive to increase production efficiency and marketable yields; the high quality standards demanded by retailers and consumers; and the increasing range of crops being grown under cover. Blue water used to produce certain types of FF&V can have very high economic returns (up to 10$/m^3 of blue water for vegetable crops grown under plastic out of season in Spain, far exceeding values for field crops), which helps explain trends towards increasing irrigation use for some types of FF&V (Fereres, Goldhamer, & Parsons, 2003).

FF&V makes a major contribution to the total blue water requirement of many diets. Fruit and vegetables (excluding potatoes) were estimated to account for 30% of the blue water required to support the diet of the UK (Hess, Andersson, Mena, & Williams, 2015). Although they exert a smaller overall impact on global water consumption than cereals due to their lower overall production levels, fruit and vegetables have the highest average water footprint per calorie compared with other crop types (with the exception of nuts and spices) (Mekonnen and Hoekstra, 2011). Additionally, the higher value of fruit and vegetable crops can stimulate greater shifts to their production in circumstances where water scarcity is driving up irrigation costs. In some instances this can have the perverse impact of driving up irrigation requirements further, as has been found for citrus in Spain (Fernandez Garcia, Rodriguez Diaz, Camacho Poyato, Montesinos, & Berbel, 2014).

Per capita FF&V consumption in the UK is among the highest in Europe, with 46% of the population consuming one to four portions per day and a third consuming more than five portions per day (Eurostat, 2018), but it has a negative FF&V trade balance and relies on imports (especially for exotics and out of season crops). In this paper we aim to identify where, geographically, the UK FF&V system is reliant on blue water resources and where it is most exposed to physical and related water risks ('hotspots'). We also identify how these have changed over the 20-year period 1996–2015.

Although there is no dataset summarizing the likelihood and severity of regulatory water risks in the form of water restrictions by production location, the degree of regional water scarcity for the production origin can be used as a proxy. Businesses that use blue water withdrawn from water scarce river basins are at a higher regulatory risk than those which use water from basins where water is plentiful (Schulte & Morrison, 2014).

This paper posits that the exposure of the UK FF&V system to physical and regulatory water risks can be inferred from the volume of blue water that is consumed in growing fruit or vegetables at a particular location and the relative abundance, or scarcity, of freshwater resources at that location. Businesses that consume large volumes of blue water in water-scarce locations are exposed to greater water risks than those that consume little or are located in basins where water is plentiful. Reputational risks are also higher in such locations as the potential damage to ecosystems and livelihoods is higher where water is scarce. The blue water scarcity footprint, WSF, is the product of the volume of blue water consumed in production and an indicator of local water scarcity. It provides an indication of the potential impact of an activity on blue water scarcity (ISO, 2014) and, from the above, it can be used as a proxy for the exposure of this point in the value chain for the product in question to physical, regulatory and reputational water risks.

Method

Supply of FF&V to the UK

In this study, FF&V includes products traded into the UK, based on those categorized in the 8-digit EU Combined Nomenclature (CN) (COEC, 1987) as Edible Vegetables (07) or Edible Fruits (08) – with the exception of products that are provisionally preserved, dried or frozen; nuts; crops grown for beverages; and crops grown for industrial processes (e.g., potatoes for manufacture of starch, grapes for wine, chillies for capsaicin or olives for oil production). The CN is based on the 6-digit Harmonised System (HS), used by the majority of trading nations, but with some categories subdivided to provide greater detail.

However, even within the CN, not all fruits are differentiated between fresh and dried (e.g., CN 08052010 does not differentiate between fresh and dried clementines). It is likely that the volumes of dried fruits in these categories that are imported are small and therefore the error introduced can be ignored. Potatoes are categorized in the CN as vegetables, but as starchy carbohydrates they may be considered to belong in a different class to other fresh vegetables. For this reason potatoes have been presented separately from other categories of vegetables in this paper.

We used UK government trade statistics (HMRC, 2017) over the period 1996 – 2015 to identify the volumes of FF&V supplied to the UK by trade, to map the quantities coming from each trading partner, and to determine how this has changed over time. Data on the annual total net mass of imports (including arrivals from the rest of the EU) to the UK from all trading partners was obtained for each specified FF&V category. The top categories were selected (see supporting information), which between them, accounted for 94.0 – 96.5% (by net mass) of FF&V traded into the UK in any year. The categories that were excluded comprized less common types of FF&V that are only supplied to the UK market in small quantities and would thus be unlikely to exert a major influence on the broad-scale water risks to the UK supply.

A proportion of the UK's FF&V supply is imported from countries that do not grow the crops themselves but are intermediaries. For example, the Republic of Ireland provides 6% of the UK's supply of bananas (HS 08039010), but given environmental production requirements they are clearly not grown there. To address this issue, re-exports to the UK from trading partners providing more than 5% (on average) of the UK's supply of an FF&V category – which for a range of types of FF&V included the Netherlands, Germany, Belgium and the Republic of Ireland – were traced back to their likely country of origin using data sourced from the United Nations Comtrade database (United Nations, 2016). Because Comtrade provides data in the 6 digit HS format rather than the 8 digit CN format it was necessary to work on the assumption that sourcing patterns would not be substantially different for the less disaggregated categories represented by the HS categories. For FF&V categories where intermediary trading partners were providing re-exported FF&V to the UK that constituted less than 5% of the UK's total supply for that category, the trade data was not traced back. Instead the overall proportional distribution of trade from producing countries for the whole dataset for that FF&V category was used as a substitute to determine the re-allocation of data from the intermediary trading port to countries of origin.

Domestic production for the relevant FF&V categories was obtained from the Department for Environment, Food and Rural Affairs' 'Horticulture statistics 2015' (DEFRA, 2016). In some cases the categorization of FF&V products did not align perfectly with the CN trade categories, however the differences in categorization were small. As such, the assumption was made that resulting inaccuracies in the analysis would not be large enough to substantially affect the outcomes of the analysis. The selected import categories and domestic production accounted for ≥ 97% of the FF&V available in the UK in any year.

Total annual exports and dispatches from the UK to the rest of the world (not broken down by trading partner countries) for each FF&V category were also obtained (HMRC, 2017). The supply of FF&V to the UK is defined as the sum of domestic production and imported produce, less the FF&V exported from the UK. To account for small year-to-year changes in sourcing, the average of 2011 – 2015 was taken as a baseline period that can be considered representative of the current sourcing of the UK's FF&V supply. To normalize for growth in supply due to increasing population, the totals were divided by the population of the UK.

Blue water consumption

Consumptive water use refers to the blue water that is removed from a drainage basin to which it is not returned, either due to evaporation, transpiration, integration into produce, or because it is released into a different basin or the sea (ISO, 2014). Where water is used but not consumed, it is returned to the same drainage basin and is thus available for other uses and users in times of water scarcity. The quantity of blue water consumed in the production of FF&V depends heavily on environmental conditions in the production location which determine the total water requirement and the need for irrigation. For example, the average blue water consumption of 1 kg of fresh potatoes grown in the UK is 11 litres (Hess, Lennard, & Daccache, 2015) whereas for production of the same quantity in Egypt an average of 290 litres is required (Mekonnen and Hoekstra, 2011). Therefore, estimates of the specific blue water consumption (l/kg) of products grown in different countries (Mekonnen and Hoekstra, 2011) have been used with the trade data to estimate the blue water consumption (*BWC*, l/year) associated with the supply of FF&V to the UK.

$$BWC = \sum_{p=1}^{p=n} \sum_{c=1}^{c=n} W_{p,c} M_{p,c}$$

Where $M_{p,c}$ is the mass (kg/year) of product, p, sourced from country, c and $W_{p,c}$ is the specific blue water consumption (l/kg) for that product and country. As Mekkonen and Hoekstra (2011) did not consider mushrooms, a specific blue water consumption of 9 l/kg was used for all locations (Robinson, Winans, Kendall, Dlott, & Dlott, 2018).

Water scarcity footprint

There are several sources of global estimates of water scarcity indicators at national or basin scales (see Liu et al., 2017 for a review). Boulay et al. (2017) present a method ('AWARE') for assessing water scarcity based on the relative available water remaining once the demand of humans and aquatic ecosystems has been met, based on modelled average runoff for 1960 – 2010. The resulting factor ranges between 0.1 and 100 and has been summarized at basin and national scale for agricultural and non-agricultural water use. A factor of 1 represents global average water scarcity whereas a factor > 1 represents increasing degrees of water scarcity. In this study, the water scarcity footprint (*WSF*, m^3 H_2O_e), in accordance with ISO (2014), is determined from

$$WSF = \frac{\sum_{p=1}^{p=n} \sum_{c=1}^{c=n} W_{p,c} M_{p,c} CF_c}{1000}$$

Where CF_c is the water scarcity characterization factor for agricultural water use in country c, (Boulay et al., 2017). The resulting water scarcity footprint is in 'water equivalent' (m^3 H_2O_e) where 1 m^3 H_2O_e can be thought of as having the same impact on water scarcity as 1 m^3 of water withdrawn from a basin at the global average level of water stress. Although the absolute numbers have no physical meaning, this provides a useful indicator for comparing the exposure of products and locations to water scarcity. A product with a large water scarcity footprint will be one where there is a coincidence of a large mass of production for the UK market, a high blue water requirement per unit of production and a water scarce location.

Small volumes of certain products are imported from countries for which no values of $W_{p,c}$ are presented in Mekonnen and Hoekstra (2011). Where a value has been given for green water consumption, it is assumed that the crop is rain fed and a value of zero has been assigned to $W_{p,c}$. Where no data are given for green or blue water it is assumed that the product is not grown in that country and has been sourced from a third country. In this case, the global average $W_{p,c}$ for that product has been used. For products sent out from the UK it has been assumed that the export crops have been domestically grown and the $W_{p,c}$ for the UK has been used, although there is likely to be some re-export of imports (e.g., an average of 20 mil kg of bananas were exported annually (2011–2015) which were not grown in the UK).

Results

Supply of FF&V

The average (2011–2015) supply of FF&V to the UK is 211 kg/capita/year, with 27% of this coming from fruit, 30% from vegetables, and 43% from potatoes (Table 1). Only 10% of the fruit is grown in the UK whilst 40% is imported from the Americas, particularly South (20%) and Central America (14%). Half of the vegetable supply is home-grown with the majority of the imported crop (41%) coming from Europe, whereas 94% of the potato supply is from domestic production. Only 3% of the FF&V available in the UK is exported, half of which is potatoes.

Blue water consumption

The net blue water consumption of the supply of FF&V to the UK is 560 million m^3/year (Table 1), 62% of which is associated with fruit production and only 26% is water

Table 1. Supply of fresh fruit and vegetables, kg/capita/year, and blue water consumption, million m^3/year, of fresh fruit and vegetable supply to the UK (2011 – 2015).

	Supply (kg/capita/year)				Blue water consumption (million m^3/year)			
Region	Fruit	Vegetables	Potatoes	Total	Fruit	Vegetables	Potatoes	Total
UK	5.8	35.4	87.2	128.4	23	38	85	146
Africa	9.7	2.0	0.2	11.9	122	9	4	135
Americas	23.3	1.4	0.0	24.8	79	4	0	83
Asia	1.2	0.9	1.2	3.3	19	3	7	29
Europe	18.2	25.7	3.7	47.6	115	63	4	182
Oceania	0.8	0.2	0.0	1.0	1	0	0	1
Source unknown	0.0	0.0	0.0	0.0	0	0	0	0
Sent out from UK	−1.8	−1.3	−2.9	−6.0	−12	−2	−3	−17
Total	57.1	64.4	89.4	211.0	347	116	97	560

consumed from domestic water resources. As such, three-quarters of the water consumed in the production of FF&V supplied to the UK is consumed in other countries, particularly Spain, South Africa and Egypt (Figure 1(b)).

Blue water scarcity footprint

The total blue water scarcity footprint of the supply of FF&V to the UK is 26,100 million m^3 H_2O_e/year. Although the UK accounts for 26% of the blue water consumption, it has much lower water stress than many of the countries from which it imports FF&V. Therefore, it accounts for only 3% of the total blue water scarcity footprint (Table 2) associated with the

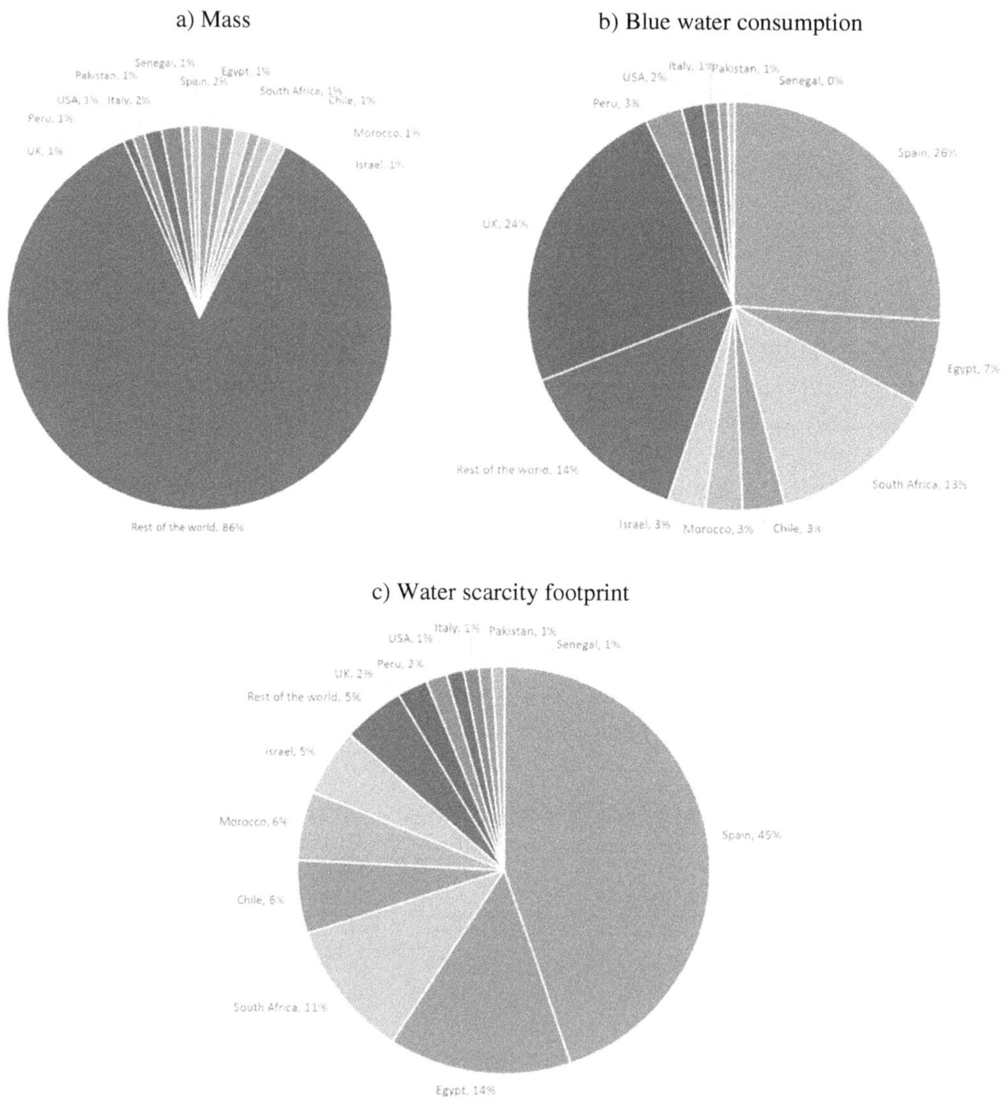

Figure 1. Average (2011 – 2015) a) sourcing of fresh fruit and vegetables to the UK, b) blue water consumption, and c) water scarcity footprint by country.

Table 2. Relative blue water scarcity footprint of fresh fruit and vegetable supply to the UK (2011 – 2015) by region.

Region	Fruit	Vegetables	Potatoes	Total
Europe	30%	17%	1%	48%
Africa	27%	3%	2%	32%
Americas	10%	1%	0%	10%
Asia	5%	1%	2%	8%
UK	0%	1%	2%	3%
Total	72%	22%	6%	100%

supply of FF&V. The largest blue water scarcity footprint is associated with fruit (30%) and vegetables (17%) from Europe and fruit from Africa (27%). Over a third of the total blue water scarcity footprint (36%) is associated with citrus (especially oranges and citrus hybrids), followed by stone fruit (10%: peaches, nectarines, plums) and tropical fruit (9%: avocados and mangoes). The vegetables with the highest blue water scarcity footprint are potatoes (6%), onions (4%) and cauliflower & broccoli (4%).

Seven countries account for 90% of the blue water scarcity footprint of the supply of FF&V to the UK (Figure 1(c)): Spain (45%: citrus, stone fruit, salads, cauliflowers & broccoli, onions), Egypt (14%: citrus, potatoes), South Africa (11%: citrus, pome fruit, stone fruit, avocados), Chile (6%: avocados, blueberries, apples), Morocco (6%: citrus, tomatoes); Israel (5%: potatoes, avocados, citrus, mangoes) and Peru (2%: mangoes, avocados, citrus) with domestic production only accounting for a further 3% (potatoes).

Changes in WSF over time

As a result of changing diet and sourcing, the per capita blue water scarcity footprint of the supply of both fresh fruit and vegetables to the UK has been increasing over the 20 year period (1996 – 2015) whilst that of potatoes has been reducing (Figure 2), primarily due to the increased substitution of potatoes by pasta and rice in the UK diet (DEFRA, 2018).

There has been a 32% increase in the per capita blue water scarcity footprint associated with fruit between 1996–2000 and 2011–2015, which can largely be explained by an increase in fruit supply (Table 3), and has resulted in increases in the blue water scarcity footprint associated with imports from Spain (citrus, stone fruit), Egypt (citrus) and Chile (avocados, blueberries) (Figure 3).

Across the vegetable category, between 1996–2000 and 2011–2015 domestic production of vegetables (excluding potatoes) has fallen by 30 kg/capita/year, which has almost entirely been compensated by increased imports from Spain and Netherlands. Less blue water is consumed in the production of vegetables in the UK (average 21.2 l/kg) and Netherlands (8.9 l/kg) compared to Spain (72.7 l/kg). In addition, although the water scarcity characterization factor for Netherlands is one-third of that of the UK, the value for Spain is 17 times that of the UK. Thus any shift to sourcing from drier and more water scarce countries results in an increase in blue water scarcity footprint. Consequently, although the supply of vegetables per capita increased by only 7%, the 62% increase in blue water scarcity footprint per capita was due to a shift in sourcing, with increases attributable to the greater reliance on vegetables sourced from Spain (salad crops, cauliflower and broccoli), Egypt (beans, onions), Senegal (sweetcorn) and Morocco (tomatoes) (Figure 4).

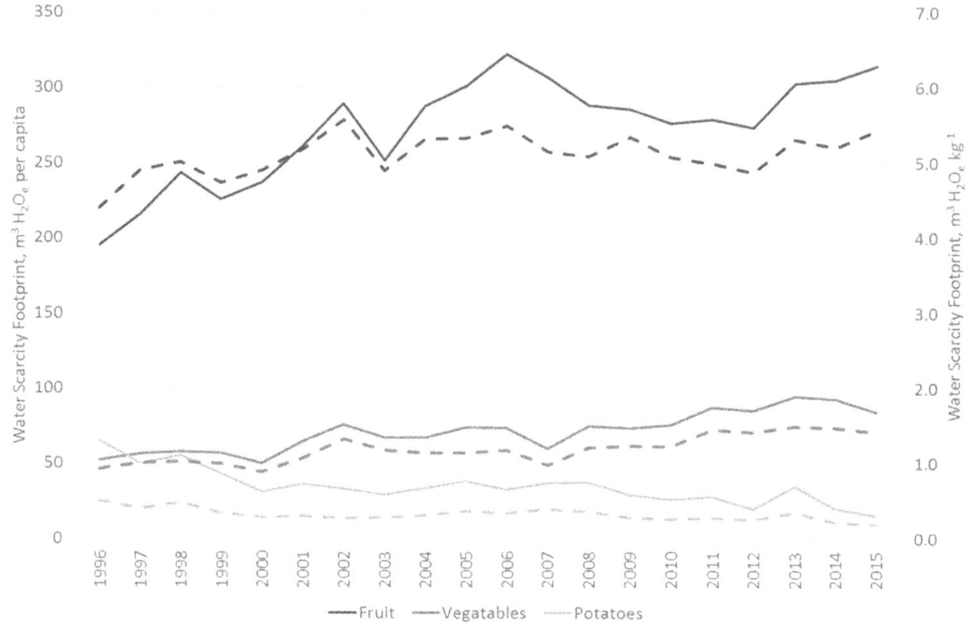

Figure 2. Change in water scarcity footprint of fresh fruit and vegetable supply to the UK (1996 – 2015) per capita (solid line) and per kg (dashed line).

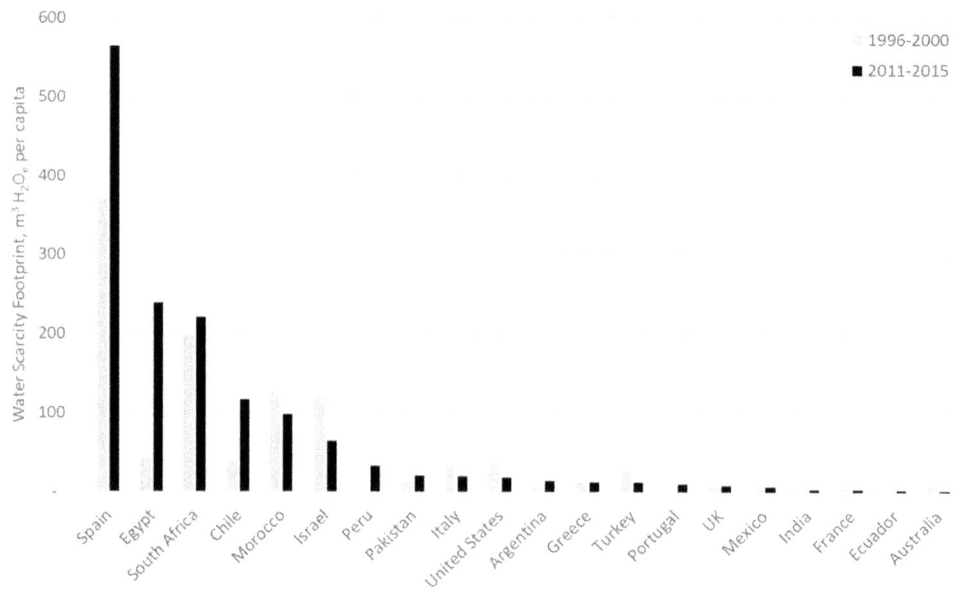

Figure 3. Change in water scarcity footprint of fruit supply to the UK between 1996–2000 and 2011–2015, m^3/capita/year showing top 20 countries by source.

In contrast, the reduced blue water scarcity footprint per capita for potatoes has been associated with a reduction in supply from 121 to 89 kg/capita/year. This has mostly come from reduced domestic production, but the fact that less is now sourced from

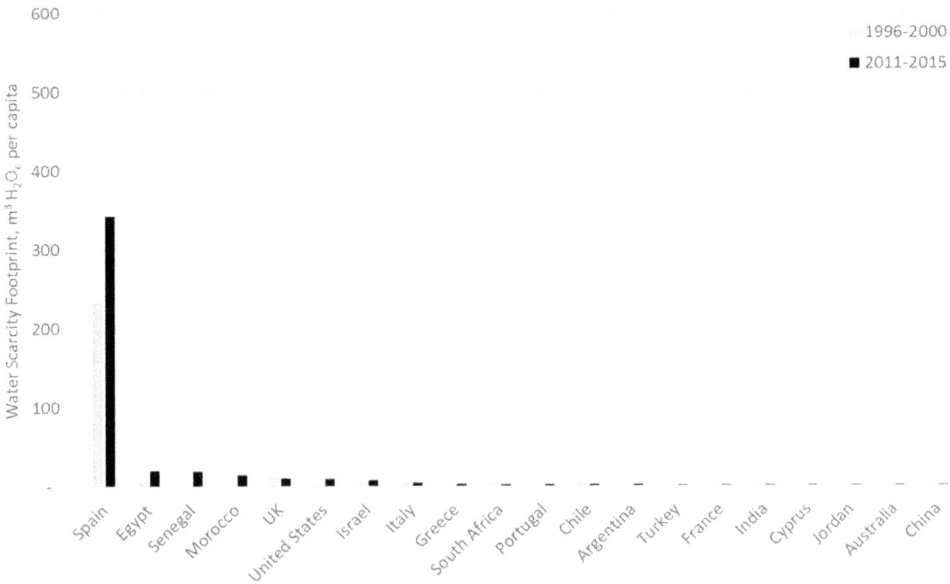

Figure 4. Change in water scarcity footprint of vegetable supply to the UK between 1996–2000 and 2011–2015, m^3/capita/year showing top 20 countries by source.

Table 3. Change in blue water scarcity footprint (m^3 H_2O_e capita/year), supply (kg/capita/year) and blue water scarcity footprint (WSF) per unit supply (m^3 H_2O_e/kg) in 2011–2015 compared to 1996–2000.

	Fruit	Vegetables	Potatoes
WSF, m^3 H_2O_e/capita/year	132%	162%	49%
Supply, kg/capita/year	123%	107%	74%
WSF, m^3 H_2O_e/kg	108%	151%	66%

Egypt, where there is a large water scarcity characterization factor, has also contributed a little to the reduction in WSF per kg.

Discussion

From this analysis, most of the UK's imports of FF&V are from water stressed countries, suggesting that the UK food system is exposed to a range of physical, regulatory and reputational water risks. Following the definition used in this study, a location with a high water scarcity characterization factor is one where the withdrawal of freshwater for agriculture has a high '... potential to deprive another user ... in this region' (Boulay et al., 2017). Therefore water for agricultural production of the UK's FF&V is in high competition with domestic, industrial and/or environmental uses.

Hotspots

Competing demands from the environment characterize the water scarcity situation for some important agricultural areas in Spain. For example, the Murcia region accounts for 20% of Spain's fruit and vegetable exports and is Spain's main producer of broccoli, celery, artichokes, apricots, table grapes, lemons and an important producer of tomatoes and peppers (Catedra UCAM-Santander, 2016). Agricultural production in the area has been intensifying in recent years, with the production of fruits increasing by 54% and vegetables by 27% between 2010 and 2015 (Catedra UCAM-Santander, 2016). Due to the extreme aridity of the local climate, production is heavily reliant on irrigation and the irrigated area has more than doubled since the middle of the last century (Oñate & Peco, 2005). Expansion of the region's canning industry coupled with severe drought in the early nineties led to both declining river flow and increasing industrial and urban wastewater discharges; a combination which resulted in severe pollution that decimated aquatic life and caused river bank degradation along large stretches of the Segura River. The public outcry that ensued led to the development of a 10 year management strategy for improving river management and significant ecological recovery has since taken place (Ródenas & Albacete, 2014). Nonetheless, continued pressure on the water resources of the region, partially attributable to production of fruit and vegetables for export, has led analysts to suggest that there is a risk of environmental and economic collapse if further new measures are not adopted to correct the effects of over-exploitation (Ruperez-Moreno et al., 2017). In this case, there is a chronic water scarcity risk to FF&V exports to the UK.

South Africa provides an example where water use for agricultural production for fruit exports is in severe competition with domestic use (Van Dam, 2017), particularly given the drought-prone nature of the region. The Western Cape province of South Africa is a major producer of export fruits for the UK, but since 2015 has been facing the combined effects of a severe drought and a strong El Niño event (Baudoin, Vogel, Nortje, & Naik, 2017). At the time of writing it is estimated to be the worst drought in over a 100 years (Van Dam, 2017). The delayed onset of winter rainfall in 2017, the high demand for water by growing urban populations and the cumulative rainfall deficit have led to the imposition of water restrictions in cities across South Africa, including Cape Town and Johannesburg (Baudoin et al., 2017; Sorensen, 2017). Farmers have been forced to reduce the irrigation of fruit orchards and take lower-yielding areas out of production altogether (Dentlinger, 2017; Jansen, 2017). Water restrictions for irrigating farmers vary from 30% to 100% in some catchments, which may have a serious impact on fruit production (Jansen, 2017). It was estimated in 2016 that the South Africa deciduous fruit industry lost ZAR720 million (≈ US$50 million) and in 2017 apple exports were down by 9% (Chambers, 2017; FruitSA, 2016). Whilst drought is a recurrent feature of South Africa's climate, its impacts on the region and the fruit sector are likely to worsen in future (Jansen, 2017), highlighting an urgent need for appropriate resilience-building measures within the FF&V value chain.

Future change

The climate in many producing regions is projected to become more unpredictable, which, when coupled with increased local demand for domestic, industrial and environmental water, is likely to lead to increased water scarcity and probability of drought. Higher temperatures will raise evapotranspiration rates and therefore irrigation demands, areas dependent on snowmelt may find the seasonal timing of water availability shifts substantially, and sea level rise coupled with increasing abstraction will likely lead to the salinization of coastal aquifers (Morison, Baker, Mullineaux, & Davies, 2008). Population growth will drive increased demand for food in the UK (as well as increased competition for food from growing and rapidly changing economies) and per capita consumption of FF&V has been increasing (Figure 2). Greater recognition of the health and environmental benefits of diets lower in processed foods, sugars and animal fats is likely to further drive up FF&V consumption in the near future, but a move to more healthy and/or vegetarian diets has been associated with an increase in blue water consumption in the UK (Hess, Andersson, et al., 2015), the USA (Tom, Fischbeck, & Hendrickson, 2016) and the EU (Vanham, Mekonnen, & Hoekstra, 2013). All these factors are likely to drive increases in the exposure of the UK's FF&V system to physical water risks in future.

Limitations and uncertainty

The estimates of specific blue water consumption for fruit and vegetable products have, necessarily, been modelled from agroclimatic data. Implicit in our assumptions is that all crops that require irrigation are actually irrigated. Mekonnen and Hoekstra (2011) concluded that when combined with national production data, and allowing for irrigation efficiency, their estimations of blue water consumption agreed well with recorded withdrawals for irrigation at national level. However, there may be large uncertainties for individual countries. For the water scarce countries that make up the hotspots of the UK's FF&V value chain, the assumption of full irrigation may be reasonable, however, in temperate climates the actual blue water consumption may be less as it may be uneconomic to irrigate crops with a small irrigation demand. For example, the estimated blue water consumption for the UK is 144 million m^3/year compared to an estimated irrigation amount of 125 million m^3/year (after Knox et al., 2015; Brown et al., 2012; and Northern Ireland Environment Agency, personal communication, 18 August 2017).

Based on the availability of trade data at the national level, water consumption and water scarcity have also been estimated at the national level. For large countries and those with diverse climates and water resource conditions this may obscure local conditions. Hess, Lennard, et al. (2015) have shown how broad-scale estimates of water scarcity footprints obscure local issues. For example the specific blue water consumption of oranges from South Africa vary between the Northern Cape province (279 l/kg) and KwaZulu-Natal province (116 l/kg) (Mekonnen and Hoekstra, 2011) whilst the Aware water stress factors are 20.1 and 5.5 respectively (Boulay et al., 2017). Thus the blue water scarcity footprint of 1 kg of oranges from Northern Cape Province is almost nine times greater than for 1 kg from KwaZulu-Natal. It was not possible to reflect these sub-national variations in water scarcity in the present study since regionally disaggregated trade data for the original production locations of the UK's FF&V supply was not available. Similarly, using an annual resolution may obscure seasonal variation in blue water demand and availability. However, whilst the trade and water scarcity information are available at a monthly resolution, the global dataset of blue water demand for all crops is only available at an annual scale. A final limitation is that, based on the sourcing data used, it is not possible to know what proportion of blue water used in each location was derived from different water sources, e.g., river, reservoirs, or groundwater, which will be exposed to differing degrees of risk.

Conclusion

13.5 billion kg of fresh fruit and vegetables are supplied to the UK each year, equivalent to 211 kg/capita/year. Of this, the UK imports 90% of the fruit and 50% of the vegetables (excluding potatoes). Growing these products consumes 560 million m^3 of freshwater per year, 74% of which is withdrawn overseas, often in countries where water resources are under pressure and drought risks are high. As a consequence, the supply of FF&V to the UK is exposed to physical, regulatory and reputational water risks in other countries as well as at home.

The locations where large volumes of water are consumed in the production of FF&V for the UK and water is scarce have the highest potential water related risks. By this logic, the UK FF&V system is particularly exposed to potential water risks in a small number of countries. Seven countries account for over 90% of the blue water scarcity footprint associated with the FF&V imported to the UK. As such, Spain (46%), Egypt (15%), South Africa (11%), Chile (6%), Morocco (6%), Israel (5%) and Peru (2%) constitute 'hotspots' where the volume of water consumed is high and water is scarce.

Over the 20 years from 1996 – 2015, the exposure of the UK's FF&V system to physical, regulatory and reputational water related risks has increased by 36% due to a combination of increased consumption of fruits that cannot be grown in the UK (especially avocados, citrus fruits and mangoes) and changes in the location of sourcing (especially a shift from domestic production to imports; predominantly from Spain). Regarding domestic production, potatoes are the product with the highest potential water related risk, but the supply of potatoes has declined dramatically over the last 20 years resulting in a lower overall water risk for domestic FF&V production. In future, growing demand for FF&V coupled with reductions in water availability (both domestic and overseas) will likely further increase the exposure of the UK's FF&V supply chain to physical, regulatory and reputational water risks and place more pressure on already stressed aquatic environments in countries that produce the bulk of the UK's supply.

There are differing potential responses to a high water risk in a FF&V supply chain. In some cases, the risk may be considered so high, that, in the long-term, alternative sourcing locations (with lower water risks) may be sought. However, this is often a last resort as resilient supply chains take a long time to establish; it may result in unacceptable social impacts on local livelihoods; and may be seen as simply offloading the problem onto others. A more sustainable approach may be to engage in catchment level-water stewardship to increase resilience to water risks. Finally, consumers may express a preference for more sustainably produced FF&V and, provided they have sufficient information to make rational choices, may prioritise certain products over others.

Whilst this global study has identified potential 'hotspots' where water risks in the FF&V value chain may be high, the coarse spatial and temporal resolution and the wide range of grower responses to water risks (e.g., through water conservation or wastewater reuse) means that local studies are needed in these areas in order to formulate water resilient supply chains.

Acknowledgments

This paper was given at the workshop on Virtual Water in Agricultural Products: Quantification, Limitations and Trade Policy (Lincoln, Nebraska, USA, 14-16 September 2016), which was sponsored by the OECD Co-operative Research Programme: Biological Resource Management for Sustainable Agricultural Systems whose financial support made it possible for the first author to participate in the workshop. This study was a re-analysis of existing data that are publicly available from HMRC at www.uktradeinfo.com, the United Nations Comtrade database at www.Comtrade.org, the UK Department for Environment, Food and Rural Affairs at www.gov.uk/government/collections/horticultural-statistics, the Water Footprint Network at www.waterfootprint.org, and the WULCA project at www.wulca-waterlca.org. The authors would like to thank David Haro-Monteagudo and Harry Hess for assistance with data processing.

Disclosure statement

No potential conflict of interest was reported by the authors.

Funding

This work was funded through the Global Food Security's 'Resilience of the UK Food System Programme', with support from the Biotechnology and Biological Sciences Research Council, the Economic and Social Research Council, the Natural Environment Research Council, and Scottish Government [Grant no. BB/N020499/1].

ORCID

Tim Hess http://orcid.org/0000-0001-5834-4269
Chloe Sutcliffe http://orcid.org/0000-0002-8035-1964

References

Baudoin, M.-A., Vogel, C., Nortje, K., & Naik, M. (2017). Living with drought in South Africa: Lessons learnt from the recent El Niño drought period. *International Journal of Disaster Risk Reduction*, *23*, 128–137.

Boulay, A.-M., Bare, J., Benini, L., Berger, M., Lathuillière, M. J., Manzardo, A., … Pfister, S. (2017). The WULCA consensus characterization model for water scarcity footprints: Assessing impacts of water consumption based on available water remaining (AWARE). *The International Journal of Life Cycle Assessment*. doi:10.1007/s11367-017-1333-8

Brown, I., Dunn, S., Matthews, K., Poggio, L., Sample, J., & Miller, D. (2012). *Mapping of water supply-demand deficits with climate change in Scotland: Land use implications, CREW report 2011/CRW006*. Retrieved September 3, 2018, from http://www.crew.ac.uk/publication/mapping-water-supply-demand-deficits-climate-change-scotland-land-use-implications

Catedra UCAM-Santander. (2016). *APROXIMACIÓN AL DIMENSIONAMIENTO DEL SISTEMA AGROALIMENTARIO DE LA REGIÓN DE MURCIA*. Murcia. Retrieved September 3, 2018, from http://www.ucam.edu/sites/default/files/universidad/informecatedra_1.pdf

Chambers, D. (2017). *Drought takes toll on apple and pear exports*. Retrieved August 24, 2017, from https://www.timeslive.co.za/news/south-africa/2017-07-10-drought-takes-toll-on-apple-and-pear-exports/

COEC. (1987). *Council regulation (EEC) no 2658/87 of 23 July 1987 on the tariff and statistical nomenclature and on the common customs tariff*. Council of the European Communities.

Retrieved September 3, 2018, from http://www.ucam.edu/sites/default/files/universidad/informecatedra_1.pdf
DEFRA. (2016). *Horticulture statistics 2015*. Department for Environment, Food & Rural Affairs. Retrieved May 9, 2018, from https://www.gov.uk/government/collections/horticultural-statistics
DEFRA. (2018). *Family food statistics*. Department for Environment, Food & Rural Affairs. Retrieved May 9, 2018, from https://www.gov.uk/government/collections/family-food-statistics
Dentlinger, L. (2017). *Western Cape declared disaster zone amid drought*. Retrieved August 24, 2017, from http://ewn.co.za/2017/05/22/western-cape-declared-disaster-zone-amid-drought
EC. (2000). *Directive 2000/60/EC of the European parliament and of the council establishing a framework for the community action in the field of water policy*. Brussels, Belgium: European Commission.
Eurostat. (2018). *The fruit and vegetable sector in the EU - a statistical overview*. Retrieved May 9, 2018, from http://ec.europa.eu/eurostat/statistics-explained/index.php/The_fruit_and_vegetable_sector_in_the_EU_-_a_statistical_overview
Falkenmark, M. (1995). Land-water linkages: A synopsis. In T. H. Mather (Ed.), *FAO land and water bulletin no. 1. Land and water integration and river basin management* (pp. 15–16). Rome, Italy: Food and Agriculture Organisation.
FAO. (2017). *AQUASTAT, FAO's information system on water and agriculture*. Rome, Italy: Food and Agriculture Organisation.
Fereres, E., Goldhamer, D. A., & Parsons, L. R. (2003). Irrigation water management of horticultural crops. *HortScience*, *38*(5), 1036–1042.
Fernandez Garcia, I., Rodriguez Diaz, J. A., Camacho Poyato, E., Montesinos, P., & Berbel, J. (2014). Effects of modernization and medium term perspectives on water and energy use in irrigation districts. *Agricultural Systems*, *131*, 56–63.
Fotheringham, A. (2012, June). Strawberry farms suck Spain dry. *The Independent*.
FruitSA. (2016). *Drought results in major losses*. Retrieved August 24, 2017, from http://www.fruitsa.co.za/2016/02/drought-results-in-major-losses/
Hess, T., Andersson, U., Mena, C., & Williams, A. (2015). The impact of healthier dietary scenarios on the global blue water scarcity footprint of food consumption in the UK. *Food Policy*, *50*, 1–10.
Hess, T. M., Lennard, A. T., & Daccache, A. (2015). Comparing local and global water scarcity information in determining the water scarcity footprint of potato cultivation in Great Britain. *Journal of Cleaner Production*, *87*(1), 666–674.
HMRC. (2017). *UK overseas trade statistics*. Retrieved September 3, 2018, from http://www.uktradeinfo.com/Statistics/OverseasTradeStatistics/Pages/OTS.aspx
Howitt, R., Medellín-Azuara, J., & Macewan, D. (2014). Economic analysis of the 2014 drought for California agriculture. *Center for Watershed Sciences. University of California, Davis, California*, 20p. Retrieved from http://watershed.ucdavis.edu
ISO. (2014). *Environmental management – Water footprint – Principles, requirements and guidelines* (Vol. ISO 14046.). Geneva, Switzerland: International Organization for Standardization.
Jansen, C. (2017). *Western Cape drought: A natural and man-made disaster*. Retrieved August 24, 2017, from http://www.freshplaza.com/article/180207/Western-Cape-drought-a-natural-and-man-made-disaster
Knox, J., Daccache, A., Hess, T., & Haro, D. (2016). Meta-analysis of climate impacts and uncertainty on crop yields in Europe. *Environmental Research Letters*, *11*(11), 113004.
Knox, J., Hess, T., Daccache, A., & Wheeler, T. (2012). Climate change impacts on crop productivity in Africa and South Asia. *Environmental Research Letters*, *7*(3), 034032.
Knox, J. W., Rickson, R. J., Weatherhead, E. K., Hess, T. M., Deeks, L. K., Truckell, I. J., … Daccache, A. (2015). *Research to develop the evidence base on soil erosion and water use in agriculture. Cranfield university for the adaptation sub-committee*. Retrieved September 3 2018, from http://www.theccc.org.uk/publication/cranfield-university-2015-for-the-adaptation-sub-

committee-research-to-develop-the-evidence-base-on-soil-erosion-and-water-use-in-agriculture/
Kumar, P., Rouphael, Y., Cardarelli, M., & Colla, G. (2017). Vegetable grafting as a tool to improve drought resistance and water use efficiency. *Frontiers in Plant Science*, *8*(June), 1130.
Lawrence, F. (2010, September). How Peru's wells are being sucked dry by British love of asparagus. *The Guardian*.
Liu, J., Yang, H., Gosling, S. N., Kummu, M., Flörke, M., Pfister, S., ... Oki, T. (2017). Water scarcity assessments in the past, present, and future. *Earth's Future*, *5*(6), 545–559.
Mekonnen, M. M., & Hoekstra, A. Y. (2011). The green, blue and grey water footprint of crops and derived crop products. *Hydrology and Earth System Sciences*, *15*(5), 1577–1600.Retrieved from
Morison, J. I., Baker, N. R., Mullineaux, P. M., & Davies, W. J. (2008). Improving water use in crop production. *Philosophical Transactions of the Royal Society B*, *363*(1491), 639–658.
Oñate, J. J., & Peco, B. (2005). Policy impact on desertification: Stakeholders' perceptions in southeast Spain. *Land Use Policy*, *22*(2), 103–114.
Ripoll, J., Urban, L., Staudt, M., Lopez-Lauri, F., Bidel, L. P. R., & Bertin, N. (2014). Water shortage and quality of fleshy fruits-making the most of the unavoidable. *Journal of Experimental Botany*, *65*(15), 4097–4117.
Robinson, B, Winans, K., Kendall, A., Dlott, J, & Dlott, F. (2018). A life cycle assessment of Agaricus bisporus mushroom production in the USA. *The International Journal of Life Cycle Assessment*. doi: 10.1007/s11367-018-1456-6
Ródenas, M. A., & Albacete, M. (2014). The River Segura: Reclaimed water, recovered river. *Journal of Water Reuse and Desalination*, *4*(1), 50.
Ruperez-Moreno, C., Senent-Aparicio, J., Martinez-Vicente, D., Garcia-Arostegui, J. L., Calvo-Rubio, F. C., & Perez-Sanchez, J. (2017). Sustainability of irrigated agriculture with over-exploited aquifers: The case of Segura basin (SE, Spain). *Agricultural Water Management*, *182*, 67–76.
Schulte, P., & Morrison, J. (2014). *The CEO water manadate. Driving harmonization of water-related terminology. Discussion paper*. The Global Compact. Retrieved on September 3, 2018, from http://ceowatermandate.org/disclosure/resources/driving/
Sorensen, P. (2017). The chronic water shortage in Cape Town and survival strategies. *International Journal of Environmental Studies*, *74*(4), 515–527.
Tom, M. S., Fischbeck, P. S., & Hendrickson, C. T. (2016). Energy use, blue water footprint, and greenhouse gas emissions for current food consumption patterns and dietary recommendations in the US. *Environment Systems and Decisions*, *36*(1), 92–103.
United Nations. (2016). *UN Comtrade Database*. Retrieved September 3, 2018, from http://comtrade.un.org/.
Van Dam, D. (2017). *Cape Town contends with worst drought in over a century*. Retrieved August 24, 2017, from http://edition.cnn.com/2017/05/31/africa/cape-town-drought/index.html
Van Vliet, M. T. H., Franssen, W. H. P., Yearsley, J. R., Ludwig, F., Haddeland, I., Lettenmaier, D. P., & Kabat, P. (2013). Global river discharge and water temperature under climate change. *Global Environmental Change*, *23*(2), 450–464.
Vanham, D., Mekonnen, M. M., & Hoekstra, A. Y. (2013). The water footprint of the EU for different diets. *Ecological Indicators*, *32*, 1–8.
Vörösmarty, C. J., McIntyre, P. B., Gessner, M. O., Dudgeon, D., Prusevich, A., Green, P., ... Davies, P. M. (2010). Global threats to human water security and river biodiversity. *Nature*, *467*(7315), 555–561.

Advising Morocco: adopting recommendations of a water footprint assessment would increase risk and impair food security for the country and its farmers

Dennis Wichelns

ABSTRACT

A water footprint assessment is not an appropriate guide for allocating water efficiently in water-scarce regions such as Morocco. Water footprints contain too little information to identify policies or investments that will achieve public goals regarding agriculture, natural resources and livelihoods. In fact, they would impair efforts to enhance food security and sustain economic growth. The better way forward for Morocco and other countries is to analyze the full range of economic, social and environmental issues pertaining to water use in agriculture and other sectors.

Introduction

Many authors in recent years have calculated the water footprints of agricultural, commercial and industrial activities, often with the goal of comparing water use within or across production settings or regions (Gu et al., 2015; Lovarelli, Bacenetti, & Fiala, 2016; Mekonnen & Hoekstra, 2014a). Within agriculture, some authors compare crop water footprints on farms (Huang, Xu, Ridoutt, & Chen, 2015; Murphy et al., 2017), while others calculate the aggregate water footprints of crop and livestock production for large regions or countries (Chen, Wu, Lei, & Li, 2017; Deng, Ma, & Li, 2016). Many of the authors suggest that water-use efficiency can be improved by reallocating water or reorganizing crop production in favour of farms, regions or countries with lower water footprints (Duarte, Pinilla, & Serrano, 2014; Mekonnen & Hoekstra, 2014a; Rudenko, Bekchanov, Djanibekov, & Lamers, 2013; Shi, Liu, & Pinter, 2014; Zoumides, Bruggeman, Hadjikakou, & Zachariadis, 2014).

Most recently, several authors have proposed using water footprints as indicators when assessing the sustainability, efficiency and equity of water allocations (Donoso, Blanco, Franco, & Lira, 2016; Hoekstra, Chapagain, & Zhang, 2016; Konar, Reimer, Hussein, & Hanasaki, 2016; Pellicer-Martínez & Martínez-Paz, 2016). Based on their calculations of regional or national water footprints, the authors recommend changes in the organization of crop and livestock production, or in public policies and investments involving agriculture, natural resources, food consumption or international trade. Many

---

authors also suggest that national or municipal governments should endeavour to motivate changes in consumer behaviour, such as reducing meat consumption, based only on their calculations of water footprints (Hoekstra, 2014; Vanham, Gawlik, & Bidoglio, 2017; Vanham, Mak, & Gawlik, 2016).

The increasing interest in calculating water footprints and developing national and regional policy recommendations based only on those calculations is somewhat surprising, given the very limited amount of information considered. Water footprints generally are calculated by dividing the amount of water applied or consumed in a production process by the amount or monetary value of the output generated (Lovarelli et al., 2016; Mekonnen & Hoekstra, 2014a). Thus, water footprints are ratios depicting the average amount of water applied or consumed per unit of output (e.g., m^3 per ton or m^3 per Euro). Because water footprints represent average calculations rather than describing incremental gains and costs, the ratios are not helpful in determining optimal strategies (Gawel & Bernsen, 2011a; Wichelns, 2015a). Smaller water footprints are not necessarily better than larger water footprints, even in water-scarce areas (Wichelns, 2011a, 2011b).

Calculations of water footprints also suffer from their consideration of only one input. Most production processes involve many inputs, some of which might be as scarce or as costly as water in a given region or production setting. Water footprints also do not account for the opportunity costs or scarcity values of water, which can vary substantially across seasons, years and locations (Gawel & Bernsen, 2013; Wichelns, 2010a, 2010b). Also missing in water footprint calculations are the likely impacts of changes in water use on the environment, livelihoods, and household health and welfare (Wichelns, 2015b, 2015c). Indeed, much of the information that matters most when evaluating policy alternatives and assessing investment options is not considered when calculating water footprints.

Although the amount of information contained in water footprints is limited, and the ratios reflect only the average amount of water applied or used per unit of output, many authors assign policy relevance to their calculations. Yet, the policy recommendations based on water footprint calculations are not supported by analysis of the broader set of variables that influence production and consumption decisions and largely determine the impacts of policy choices on individuals, households and communities. Such variables include the prices and availability of resources other than water, interactions involving water and other inputs, market access, and the risk and uncertainty associated with household and regional production decisions, particularly in agricultural settings.

Pahlow, Snowball, and Fraser (2015) calculate water footprints for food and biofuel crops in South Africa. They suggest that the government should implement incentives to motivate the import of water-intensive crops to alleviate water scarcity, while also encouraging residents to consume less meat. However, the authors provide no analysis of food, water or fuel demands in South Africa, in a manner that would enable policymakers to assess the likely implications of incentives regarding imports and exports on commodity prices, farm incomes and household welfare. The recommendation regarding meat consumption also lacks support from any analysis of the likely economic, cultural or health implications of a notable change in food consumption patterns.

Zoumides et al. (2014) offer a similar recommendation, based on their calculations of national water footprints in Cyprus. They suggest that the government should encourage consumers to reduce water demands, partly by adopting diets involving less meat. The authors assume that such a shift in consumption would allow more water to be used for food crop production, but they provide no analysis supporting such an outcome. In a similar fashion, Schyns, Hamaideh, Hoekstra, Mekonnen, and Schyns (2015) suggest that the government of Jordan should endeavour to reduce water demand in the country, in part by motivating consumers to reduce meat consumption. The authors recommend implementing meat-free days and labelling products to reflect their calculated water footprints. No analysis regarding the likely impacts of such measures is provided.

Mekonnen and Hoekstra (2014b) suggest that Kenya should reduce water scarcity in the country by increasing the import of water-intensive products, such as cereals, and increasing the export of high-value products, such as cut flowers, vegetables and spices. Such a recommendation might seem compelling at first glance, but the authors examine no information regarding production costs, market demands or the endowments of other agricultural inputs at the farm level or in aggregate. Chouchane, Hoekstra, Krol, and Mekonnen (2015) offer a similar suggestion based on their calculation of water footprints in Tunisia. They suggest that the government should encourage farmers to switch from grains to higher valued crops in areas where market conditions are supportive and grains can be imported. Yet, the authors offer no analysis of import and export markets, agricultural demand and supply, or the welfare effects of current and proposed policies pertaining to international trade. In both Kenya and Tunisia, the limited information contained in the calculated water footprints simply is not adequate to reach decisions regarding policies that will influence production choices and trading strategies.

The goal in the present paper is to demonstrate empirically that policy recommendations based on calculations of water footprints are misplaced, inappropriate and potentially harmful. To this end, the policy recommendations arising from a water footprint assessment performed in Morocco are considered (Schyns & Hoekstra, 2014a). Working with the empirical information provided in that assessment, it is shown that agricultural production, farm incomes and food security likely would be impaired across the country if the Moroccan government adopted the recommendations put forth by the authors. Public officials wishing to improve productivity, enhance livelihoods and ensure food security in Morocco and elsewhere must consider a much broader set of information than that which is contained in calculations of water footprints.

Assessing water footprints in Morocco

Schyns and Hoekstra (2014a) suggest that Morocco's national water policy can be enhanced by calculating and comparing water footprints in agriculture, industry and domestic water supply. They calculate crop water footprints in each of Morocco's river basins. The authors also calculate the volumes of 'virtual water' used to produce the goods and services traded between Morocco and other countries. The calculations of water footprints and virtual water volumes are used to identify response options that will 'alleviate water scarcity and allocate water resources more efficiently' in Morocco (Schyns & Hoekstra, 2014a, p. 2). Among the options the authors identify is the relocation of crop production activities across river basins, in the interest of reducing the amount of water consumed in agriculture. To that end, they propose relocating all maize production in Morocco to the Moulouya River Basin, thus increasing the area planted in maize in that basin from an average of 838 ha during 1996–2005 to more than 220,000 ha in future.

Schyns and Hoekstra (2014a) also recommend relocating the production of wheat and barley, which provide much of Morocco's daily supply of calories and protein. They suggest the relocation programme will save substantial water volumes, yet they do not consider the costs or risks involved in such a substantial shift in agricultural activities. Nor do they discuss the potentially negative impacts on national and household food security of a programme that would greatly reduce crop diversification across regions, thus increasing the likelihood of major crop losses due to drought, pests, and the changes in temperature and rainfall that likely will accompany climate change (Droogers et al., 2012; Schilling, Frier, Hertig, & Scheffran, 2012; Terink, Immerzeel, & Droogers, 2013). Rather than adding value to national water policy, the recommendations arising from the water footprint assessment, if implemented, could greatly impair national efforts to sustain economic development and ensure food security.

The Schyns and Hoekstra (2014a) study of water footprints in Morocco demonstrates quite well that policy-relevant insight cannot be obtained from calculations involving only aggregate crop production data and estimates of water use. Calculations of water footprints cannot capture the complexity of agriculture or reflect the range of issues that must be considered when determining national strategies regarding natural resources. Water footprints do not describe the timing, quality or amounts of other inputs used in crop production, such as seeds, plant nutrients, pesticides, labour and capital. A water footprint assessment does not account explicitly for the inherent variation in crop water use and yields, over time and across locations. Understanding that variation is critical in determining optimal policy choices and selecting investments to maximize and sustain the benefits generated in agriculture.

Assessing water footprints in the interest of guiding national water policy is particularly misplaced in rainfed settings, where farmers have little control over the amount or timing of soil moisture available each season. Crop yields and water use vary substantially in such settings, such that farmers must focus on managing risk and uncertainty, rather than minimizing water footprints (Magnan, Lybbert, Mrabet, & Fadlaoui, 2011; Wichelns, 2015a). Given that 85% of Morocco's farmland is rainfed rather than irrigated, national policies that assist farmers in reducing risk and uncertainty, and investments that increase crop yields while reducing annual variation, will be most appropriate. Indeed, throughout much of west Asia and North Africa, investments that enhance crop yields in rainfed settings will be most effective in ensuring national and household food security (Haddad et al., 2011).

The analysis that follows describes the notable risk and limited productivity of agriculture in Morocco and considers the country's objectives regarding food security and economic development, while demonstrating the inappropriateness of a water footprint assessment. Agriculture in Morocco is largely rainfed, as only 10–15% of the agricultural area is irrigated (Jacobsen, Jensen, & Liu, 2012; Lyamani, Silué, & Guéi, 2011). Given the sparse and uncertain rainfall in many areas of the country, crop yields are small and variable (Mrabet, Moussadek, Fadlaoui, & Van Ranst, 2012; Pala et al., 2011). Most farmers, particularly smallholders, plant cereals to minimize the risk of financial losses in dry years. Large, corporate farms, with access to irrigation, are better placed to invest in fruit and vegetable production. Efforts to motivate improvements in water management or increase agricultural productivity must acknowledge the inherent stochastics of crop production and the very different production opportunities and constraints on small and large farms in Morocco.

Policy analysts must also acknowledge the importance of sustaining economic growth, increasing employment opportunities, reducing poverty, and ensuring national and household food security. Agriculture is the primary source of income for much of Morocco's population, particularly in rural areas. Major changes in crop and livestock production patterns could have substantial, negative impacts on millions of residents who have very few alternative livelihood opportunities. In addition, international trade in agricultural products is a critical component of Morocco's national food security. Any effort to modify the country's trade strategy, based on crop water requirements, could reduce the generation of foreign exchange, upset the financial balance between imports and exports, and greatly diminish the social benefits of engaging in international trade.

Agriculture is complex and inherently stochastic

Agriculture accounts for about 13–20% of gross domestic product (GDP) in Morocco, depending on annual output, while employing about 40% of the population (Ghanem, 2015; Mrabet et al., 2012). Most farmers rely largely on rainfall, given the limited extent of irrigation in Morocco (Jacobsen et al., 2012; Lyamani et al., 2011). Cereal crops are produced on about 5 million ha (about 65% of the planted area), while generating about 55% of the total value added in agriculture (Karrou & Oweis, 2014). Cereals provide much of the country's food security, accounting on average for about 58% of the calories and 60% of the protein consumed (FAO, 2016a). Cereal production varies across provinces and over time, with differences in annual rainfall. Drought occurs frequently in Morocco, with notable impacts on cereal yields and aggregate production (Karrou & Oweis, 2014). The country relies on cereal imports for about 40% of annual consumption, although the proportion rises and falls with changes in domestic production (FAO, 2016a).

Variability is a defining characteristic of cereal production in Morocco, particularly in rainfed areas. In a study of cereal yields and rainfall in Morocco from 1988 to 2008, Karrou and Oweis (2014) show that yields varied between 150 and 3000 kg/ha, with coefficients of variation ranging from 30% to 50% in the north and from 60% to 70% in the south. The authors show also that a moderate drought occurs in Morocco once every 2.6 years, while a severe drought occurs once every seven years. Rainfall also varies spatially. The average annual rainfall varies from 1800 mm in northern Morocco to less than 200 mm in the south. In 1994, six of the 11 hydrologic basins in Morocco received less than 50% of the long-term average rainfall (Doukkali, 2005).

Wheat and barley yields in Morocco are highly correlated with the amount of rainfall received from January to March. Rainfall deficits early in the year have a notable impact on annual cereal production. The national mean wheat yield increased from about 450 kg/ha in 1961 to 1900 kg/ha in 1991, with notable variation along an upward-sloping linear trend line (Balaghi, Tychon, Eerens, & Jlibene, 2008). The upward trend is less evident since the early 1990s, while the variance has notably increased (Figure 1).

Average wheat production has moved sharply higher and lower in some years, ranging between 500 and 2200 kg/ha. Given such variation, the mean wheat yield, across years, is not a useful descriptor of wheat production potential or food security outcomes. Information describing the variation in crop yields is essential, particularly when analyzing policy options pertaining to land and water resources, farm incomes and food security.

A second defining characteristic of cereal production in Morocco is the low yield achieved, particularly in rainfed settings. In a study of wheat yields in Morocco from 1995 to 2004, Pala et al. (2011) find that wheat yields achieved by farmers vary with annual rainfall and are substantially lower than wheat yields observed on experiment stations. The experiment station yields, obtained with improved management practices, exceeded farm-level yields by 61–153%. The authors suggest that farm-level yields in Morocco could be increased by 1.6–2.5 times, with higher yielding crop varieties, improvements in soil and water management practices, greater use of fertilizer, and supplemental irrigation.

Crop yields are less variable in irrigated settings, yet the yield gap between irrigated farm-level yields and those obtained on experiment stations is substantial. Many farmers in both rainfed and irrigated settings use insufficient amounts of fertilizer and pesticides, and many apply inputs at the wrong time, in both wet and dry years (Pala et al., 2011; Ryan et al., 2012; Ryan, Ibrikci, Sommer, & McNeill, 2009). During the notably dry year of 2007–08, Abderrazzak, Daoui, Kajji, Dahan, and Ibriz (2013) obtained a 91% increase in wheat yield on experimental plots by providing 60 mm of supplemental irrigation. Benlhabib, Yazar, Qadir, Louren, and Jacobsen (2014) achieved similar yield increases in barley, wheat and quinoa, with supplemental irrigation and nutrient management, in a four-year study conducted on farm fields in Bouchane and Mnabha, Morocco.

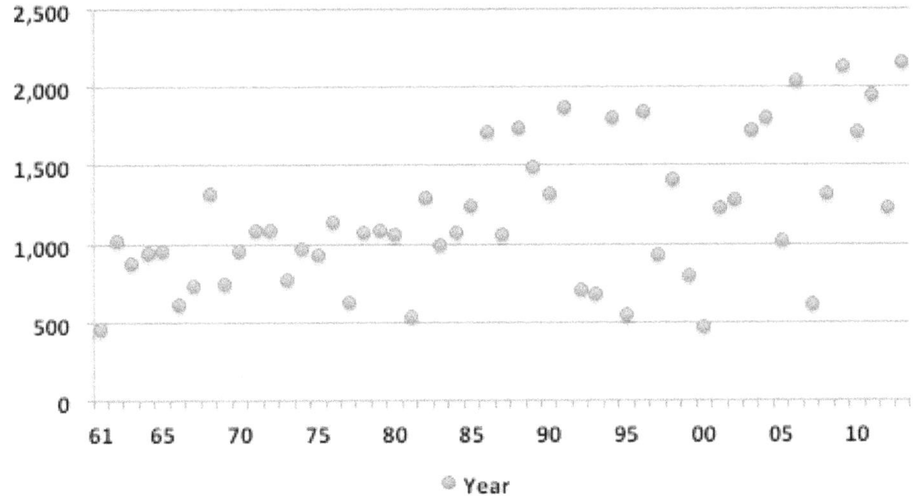

Figure 1. National average wheat production in Morocco (kg/ha), 1961–2013.
Source: FAO (2016e).

A third defining characteristic of Moroccan agriculture is the difference in the distribution of production activities between large and small holders of land. While 75% of farmland in Morocco is controlled by corporations and other large-scale operators, family farms account for about 70% of land holdings, most of which are smaller than 5 ha (Ghanem, 2015). The average size of family farm holdings in Morocco is about 2 ha. Many of the small farms produce crops for home consumption and for sale in local markets. About 10% of Morocco's population of 32 million is engaged in family farming, which provides an important source of employment in rural areas (Ghanem, 2015). Poverty is largely a rural phenomenon in Morocco, where 43% of the population lives in the countryside (Ghanem, 2015). Thus, policies and investments that influence agricultural productivity can have a notable impact on the livelihoods and food security of a substantial portion of Morocco's population.

Sustaining economic development and ensuring food security

Morocco has achieved notable economic growth in recent years, while also reducing poverty and enhancing food security. The population increased by 62% from 1980 to 2011, yet GDP increased by a factor of 3.53, such that GDP per capita increased by a factor of 2.16. During those years, even with the 62% increase in population, the average daily consumption of calories and protein per capita increased by 26% and 35%, respectively (Table 1). These gains in daily consumption were achieved even as the amount of arable land per capita declined by 35% from 1980 to 2011. Poverty and food insecurity are not yet eliminated in Morocco, but the progress achieved since 1980 is notable given the country's limited land and water resources, its reliance largely on rainfed agriculture, and the frequency with which droughts occur.

Much of Morocco's success in sustaining economic growth and enhancing food security is due to the country's investments in irrigation schemes, large dams and other infrastructure, its financial support for agriculture, and its engagement in international trade (Mahe et al., 2013; Schilling et al., 2012; Snoussi, Haïda, & Imassi, 2002). Regarding water resources, the number of large dams in Morocco has increased from 30 in 1981, with a combined capacity of 9.3 billion m^3, to 128 in 2012, with a combined capacity of 17.0 billion m^3 (Mahe et al., 2013). Additional investments in irrigation capacity and water-use efficiency are planned in future, in conjunction with the Green

Table 1. Growth ratios for population, gross domestic product (GDP), arable land and daily consumption of calories and protein in Morocco, 1980–2011.

Period	Population	GDP	GDP per capita	Arable land per capita	Calories/day	Protein/day
1996–2005	1.11	1.38	1.24	0.81	1.04	1.06
2005–2011	1.06	1.33	1.25	0.92	1.04	1.08
1980–2011	1.62	3.53	2.16	0.65	1.26	1.35

Note: Ratios are formed by dividing the value in the ending year of each period by the value in the beginning year.
Source: Population and food consumption data are from the Food and Agriculture Organization's (FAO) food balance sheets for Morocco (FAO, 2016a). The GDP and arable land data are from the World DataBank data set (World Bank, 2015).

Morocco Plan (Jobbins, Kalpakian, Chriyaa, Legrouri, & El Mzouri, 2015; Mohamed, 2013).

Agricultural production in Morocco is not sufficient to meet national food requirements, and thus the country imports substantial amounts of wheat, sugar, maize and soybean oil (Table 2). The country exports crops and processed agricultural products for which it has a comparative advantage with trading partners, including tomatoes, citrus fruit and vegetables (Table 2). The imports and exports of agricultural products vary over time, with differences in rainfall and domestic production, and with changes in market conditions. Over time, the imports of cereals and the exports of higher valued crops have increased in Morocco, with increases in population and economic growth (FAO, 2015b).

Cereal imports are particularly important in ensuring food security in Morocco, as they reduce the variance in the amount of cereals available for consumption each year. The annual production of wheat ranged from 1.4 to 6.4 million tons/year in Morocco during 1996–2011, while imports of wheat ranged from 1.8 to 4.1 million tons (Table 3). The coefficient of variation for wheat production in Morocco is 0.41, while that for wheat imports is 0.24. As a result, the coefficient of variation for the amount of wheat available per capita is 0.20, which is about half that for wheat production per capita (Table 3). Wheat imports thus reduce variability in national food security.

Morocco's cereal dependency ratio, defined as the ratio of imported cereal to total cereal consumption, is about 54%, which is notably higher than the world average ratio of 16% (Ghanem, 2015). Morocco spends about 20% of its export revenues on food imports, a ratio that also is substantially higher than the world average (Ghanem, 2015). While the 20% ratio likely can be sustained, Morocco is vulnerable to the potentially negative impacts of periodic grain shortages and price spikes on international markets, such as those that have occurred in recent years. The Green Morocco Plan is designed to increase the

Table 2. Top 10 agricultural imports and exports by value for Morocco, 2011.

	Amount (tons)	Value (US$, thousands)	Unit value (US$/ton)
Imports			
1. Wheat	3,661,864	1,322,600	361
2. Sugar (raw)	851,814	544,344	639
3. Maize	1,545,285	490,298	317
4. Soybean oil	350,748	484,531	1381
5. Cake of soybeans	551,202	247,479	449
6. Tea	64,541	219,619	3403
7. Cotton lint	34,982	112,428	3214
8. Butter (cow milk)	19,712	103,270	5239
9. Barley	299,877	93,704	309
10. Dates	33,140	75,208	2269
Exports			
1. Tomatoes	392,435	378,593	965
2. Tangerines	348,509	353,175	1013
3. Olives (preserved)	85,328	131,313	1539
4. Oranges	171,615	123,534	720
5. Chillies, peppers	61,539	106,461	1730
6. Fruit (prepared)	67,877	95,016	1400
7. Olive oil	36,004	83,313	2314
8. Strawberries	24,327	78,761	3238
9. Melons	52,683	61,954	1176
10. Cheese (processed)	6628	33,908	5116

Source: Food and Agriculture Organization's (FAO) FAOSTAT data (FAO, 2016b).

Table 3. Summary statistics describing the production and imports of wheat for Morocco, 1996–2011.

Summary statistic	Production	Imports	Sum of production and imports
Aggregate (million tons)			
Mean	4.08	2.95	7.04
Minimum	1.38	1.83	4.37
Maximum	6.37	4.09	9.68
Standard deviation	1.66	0.71	1.54
Coefficient of variation	0.41	0.24	0.22
Per capita (kg per person)			
Mean	136	99	235
Minimum	48	60	158
Maximum	217	132	302
Standard deviation	54	21	47
Coefficient of variation	0.40	0.22	0.20

Note: The coefficient of variation is calculated as: standard deviation/mean.
Source: Food and Agriculture Organization's (FAO) FAOSTAT database (FAO, 2016b).

production of higher valued crops, in part to generate additional foreign exchange to support food imports. Morocco likely will continue importing large amounts of food grains in future, given the scarcity of arable land per capita, the projections of higher average temperatures and lower annual rainfall in future, due to climate change, and the high cost of constructing additional water storage facilities (AfDB, 2012; Hellegers, Immerzeel, & Droogers, 2013; Tanasijevic, Todorovic, Pereira, Pizzigalli, & Lionello, 2014).

Shortcomings of the water footprint assessment

Given the persistent water scarcity in Morocco and the large amounts of water consumed in agriculture, estimates of crop water requirements across river basins can provide an interesting description of water use in an important sector. However, that information alone is not sufficient for guiding policy decisions. The observed patterns of crop and livestock production and water use in Morocco have evolved over many years and reflect many farm-level and national considerations involving climate, soils, technology, human capital, markets, relative prices, public policies, investments in infrastructure, and both national and household food security objectives. The patterns also reflect the opportunity costs of farm-level choices, and the risk premia that farmers place on decisions regarding higher and lower valued production alternatives. All this information will – and should – guide future production decisions. Yet, none of this information is contained in calculations of water footprints.

Given the context of advising Morocco regarding its national water policy, one might suggest there are four primary shortcomings in the water footprint assessment presented by Schyns and Hoekstra (2014a):

- The objectives of the analysis are not appropriate.
- The analytical framework is not correct.
- The interpretation of the calculated ratios is not accurate.
- The optimization analysis is misguided.

As a result, the policy recommendations arising from the water footprint assessment are misplaced. Each of these shortcomings is described in order, relying largely on information provided by Schyns and Hoekstra (2014a, 2014b), which describes in greater detail the methodology and data used in the water footprint assessment.

The objectives of a water footprint assessment are not appropriate

Schyns and Hoekstra (2014a, p. 2) assess the water footprint of activities in Morocco and the 'virtual water trade balance' of the country, with the goal of identifying response options that 'will reduce the water footprint within Morocco, alleviate water scarcity, and allocate water resources more efficiently'. This is a tall order for analysis involving a ratio that includes only an estimate of water use and some measure of output (crops, livestock or industrial production). In addition, neither Morocco nor any of its trading partners engages in the trade of crop or livestock water requirements. The country trades in crop and livestock products. Thus, the notion of 'virtual water trade' does not describe the actual goals, gains or firm-level decisions pertaining to international trade (Gawel, 2014; Wichelns, 2010a, 2010b).

Regarding efficiency, water footprints are not an appropriate metric, as the ratios represent averages rather than incremental rates of change, which are required to determine efficient outcomes. In addition, water footprints do not account for the opportunity costs of water use in any setting. Thus, the ratios are not helpful in determining the optimal allocation of water resources (Gawel & Bernsen, 2011a; Wichelns, 2015b). Furthermore, only one-third of Morocco's renewable water supply can be allocated in the form of withdrawals from surface water or groundwater sources. Most of the renewable supply is transpired by crops and native vegetation in rainfed settings. In particular, of the 29 km^3 long-term average renewable freshwater supply, an estimated 10.35 km^3 (35.7%) are withdrawn for agricultural, municipal and industrial uses (FAO, 2016c). Of that volume, 9.16 km^3 (31.6% of the long-term average) are withdrawn for agriculture. Opportunities to reallocate the 10.35 km^3 of water withdrawals among competing uses will be limited by any pre-existing claims to water resources, any constraints imposed by the lay of the land, and by the extent and capacity of water delivery facilities. In any event, a comparison of water footprints will not reveal the optimal allocation.

The goal of alleviating water scarcity might be appropriate in local or subnational settings within Morocco, yet in aggregate, water will remain quite scarce for the foreseeable future. Given the long-term average renewable freshwater supply of 29 km^3, and a current population of about 34 million, the estimated average freshwater availability per capita is 844 m^3 (Table 4). That volume is within the range of 500–1000 m^3 per capita considered by some observers to represent a condition of chronic water scarcity (Tekken & Kropp, 2012). The current population growth rate of about 1.3% per year is projected to decline to 0.7% in 2030 and 0.3% in 2050. Thus, the population will continue increasing in Morocco, reaching 39.8 million in 2030 and 43.7 million in 2050 (Table 4). In those years, the average freshwater availability per capita will decline to 729 and 664 m^3 respectively (Table 4). Morocco can build additional storage to capture and store a larger portion of its annual rainfall, and firms and consumers can improve their water management practices, yet aggregate water scarcity will increase in future with increases in population and continued economic growth. In addition,

Table 4. Population, annual growth rate and average freshwater resource available per capita in Morocco, 2015–50.

Year	Population (thousands)	Population growth rate (%)	Available freshwater (m^3 per capita)
2015	34,378	1.3	844
2030	39,787	0.7	729
2050	43,696	0.3	664

Note: The long-term average renewable freshwater supply is 29 billion m^3 (FAO, 2016c).
Source: Population and growth rate data are from the Food and Agriculture Organization's (FAO) FAOSTAT database (FAO, 2016d).

climate change is expected to bring an increase in average temperatures and a reduction in the average annual rainfall (Saadi et al., 2015; Schilling et al., 2012; Terink et al., 2013).

The analytical framework is not correct

Often implicit in the discussion of a water footprint assessment is the notion that a smaller water footprint generally is better than a larger one. Indeed, Schyns and Hoekstra (2014a) describe in their introduction the goal of identifying policy options that will reduce the water footprint in Morocco. Yet, there is no inherent, conceptual basis for seeking or achieving such an outcome. Water is scarce in Morocco, and it should certainly be used efficiently. However, arable land, energy, skilled labour and other inputs also are scarce, and each must also be allocated and used in a manner that maximizes some measure of net benefits. It is not possible to determine the optimal allocation of water or any one resource without also considering the roles and scarcity values of other inputs in production. In many settings, the incremental cost of reducing a water footprint might exceed the incremental gain, particularly if the effort requires a large expenditure on energy or a large investment in facilities or equipment.

Water footprints are ratios describing the average amount of output generated (in mass or value) per some measure of water input (water applied or consumed). As such, the ratios do not describe incremental gains or costs, and thus cannot be used to identify economically optimal strategies (Gawel & Bernsen, 2011a; Wichelns, 2015b). Water footprints also fail to consider the opportunity costs of water and other inputs (Gawel & Bernsen, 2011a; Wichelns, 2015b). The notion of a water footprint is not an optimizing criterion. It is merely a descriptive ratio that includes very limited information.

In addition to falling short as an optimizing criterion, water footprints do not correctly characterize water use. Perry (2014) has shown that the water footprint perspective fails to acknowledge inherent complexities in describing and managing water resources, and that estimates of water footprints do not account correctly for evapotranspiration (ET). Many of the authors who present estimates of water footprints, including Schyns and Hoekstra (2014a), rely on the estimation procedure described in *The Water Footprint Assessment Manual* (Hoekstra, Chapagain, Aldaya, & Mekonnen, 2009, 2011). Perry (2014) has identified several shortcomings of that procedure, particularly when applied in irrigated settings:

- The assumption that crop ET from any source is equal to potential crop ET is incorrect when crop yields are less than potential. When plants are stressed, transpiration slows, such that actual ET will be less than potential ET. As a result, estimated water footprints generally overstate actual ET.
- Because actual ET generally is less than potential ET, while the contribution of rainfall is fixed, any overestimation of ET is assigned to irrigation.
- The sources and uses of moisture in the soil before and after the crop season affect estimates of the contribution of soil moisture and irrigation to total consumption.

In sum, water footprints are both conceptually and empirically unsuitable for identifying desirable outcomes or determining optimal strategies. Those tasks require economic analysis that includes the consideration of a wide range of inputs, multiple objectives and opportunity costs.

There is also no conceptual or empirical support for deriving policy implications based on the 'flows of virtual water' between countries. Schyns and Hoekstra (2014a) suggest that Morocco imports and exports 'virtual water', and they calculate the average US dollar values per unit of 'virtual water' imported and exported. The authors also report that about 4% of the water used in agriculture and industry in Morocco is used to produce goods that are exported. Such calculations might be of interest to some observers, but they provide no insight regarding the costs or benefits of engaging in international trade. There is no conceptual basis for assessing the gains from trade by dividing the value of imports and exports by the units of a single input used in production (Wichelns, 2004, 2011a, 2011b). The prescriptions of virtual water analysis regarding international trade do not provide useful guidance for firms or policy-makers. Rather, they can result in misleading and inefficient recommendations (Gawel & Bernsen, 2011a). Even in water-scarce regions, countries must consider much more than their water endowment when selecting viable paths of economic development (Gawel & Bernsen, 2013).

In their optimization analysis pertaining to cropping patterns, Schyns and Hoekstra (2014a, p. 5) introduce the notion of a crop-specific 'benchmark water footprint', which they define as the 'lowest water consumption [...] of that crop, which is achieved in a comparable river basin in Morocco'. They then assess the potential water savings that could be achieved by reducing the water footprints of 12 major crops to their benchmark water footprints. That analysis might seem compelling at first glance, yet the notion of a benchmark water footprint is inappropriate for two reasons:

- The authors calculate average water footprints over a 10-year period, during which rainfall, crop yields and water use vary substantially. Calculations depicting averages of those variables are not helpful in determining strategies for optimizing crop production in an environment with extraordinary inherent variation.
- There is no conceptual or empirical basis suggesting that the farm-level or national goal should be that of minimizing water footprints. It is more likely that policy-makers and farmers are focused more intently on minimizing the potential harm due to substantial variation in annual rainfall and crop yields, within the larger context of sustaining economic growth and ensuring national and household food security.

This rationale also applies to the authors' analysis in which cropping patterns are modified by relocating crops to river basins where the observed water footprints are smallest. As described above, water footprints contain too little information to guide policy decisions. This inadequacy is magnified when suggesting that a water footprint calculated as the average ratio for a 10-year period in a highly variable setting should be used as the criterion for relocating crop production across river basins.

The interpretations are inaccurate

Schyns and Hoekstra (2014a, p. 3) describe their work as a study of 'economic land and water productivities' in Morocco. They interpret their results as suggesting that 'from a strictly water-economics point of view, it would be worthwhile to reconsider which crops to grow in Morocco (due to the low value in US$/m^3 and US$/ha for some crops compared to others)' (p. 10). This perspective is misplaced given that the authors have not conducted an economic analysis of land or water use in Morocco. They have merely calculated ratios depicting the average amount of output associated with the average amount of a single input. As noted above, economic analysis requires a consideration of incremental values, multiple inputs and objectives, and opportunity costs.

Schyns and Hoekstra (2014a, p. 5) report that the five crops using most water in Morocco are those with 'the lowest economic water productivity, ranging from 0.08 US$/m^3 for wheat to only 0.02 US$/m^3 for almonds'. Although these average calculations do not represent economic productivity, the low values partly reflect the challenges of producing crops in rainfed settings in which the risk of crop failure prevents farmers from optimizing the use of all inputs. Even in irrigated settings, the average and incremental returns to water in production often are in the range of the values observed by Schyns and Hoekstra.

Ziolkowska (2015), in a study of irrigated crop production in the Texas High Plains region of the United States, reports shadow prices of water (the incremental value that a farmer could afford to pay) in 2010 that include 0.07 US$/m^3 for maize, 0.02 US$/m^3 for soybeans and 0.05 US$/m^3 for cotton.

Frija et al. (2014) examine irrigated wheat production in central Tunisia, where farmers purchase irrigation water for 0.08 US$/m^3. The authors estimate a production function and determine that the optimal water application is 2700 m^3/ha. In their sample of 150 randomly selected farmers, 45% apply less than 2500 m^3/ha of irrigation water, while 31% apply more than 3500 m^3/ha. Thus, the observed incremental value of irrigation water varies substantially around the optimal value of 0.08 US$/m^3.

Schyns and Hoekstra (2014a, p. 4) suggest that they 'assess the environmental sustainability of the water footprint within Morocco', in part by comparing 'the water needed to assimilate the nitrogen fertilizers that reach the water systems due to leaching [...] with the waste assimilation capacity of aquifers' (p. 4). They define water pollution as 'the total grey water footprint in a catchment divided by the waste assimilation capacity' (p. 4). A ratio > 1 indicates that 'ambient water quality standards are violated' (p. 4). The notion of a grey water footprint is the volume of water required to dilute a pollutant to the concentration described in a prevailing ambient water quality standard (Franke, Boyacioglu, & Hoekstra, 2013). In their empirical analysis, Schyns and Hoekstra (2014a) compare their estimates of grey water footprints with their estimates of groundwater assimilative capacity. They interpret their results as indicating that water quality standards are violated in three river basins, while not in six others. They suggest also that the water pollution level in the Bouregreg basin is infinite.

The constructs of a grey water footprint and a water pollution ratio are contrived in a fashion that does not provide insight regarding actual water pollution issues in Morocco or elsewhere. Characterizing water pollution by estimating the amount of dilution water required to achieve an ambient concentration standard pertaining to one constituent (nitrate) provides no information regarding the impacts of nitrate or other pollutants on the environment. Long ago, most specialists dismissed the notion that dilution is an appropriate strategy for alleviating the impacts of pollution. Source control and water treatment are the remedies imposed in most jurisdictions to protect and improve water quality. Key issues for policy-makers include the incremental costs and benefits of such efforts and the incentives, regulations and monitoring programmes required to achieve compliance.

Water quality issues in Morocco and elsewhere involve considerations that cannot be addressed by calculating grey water footprints. Re et al. (2013) describe the impacts of nitrates from agriculture and in effluent from urban settlements on groundwater quality in the Bou-Areg coastal aquifer in northern Morocco. Increasing salinization of the aquifer also is of concern, due partly to both natural weathering processes and salts in agricultural return flows (El Yaouti, El Mandour, Khattach, Benavente, & Kaufmann, 2009). Re et al. (2013) also describe the impacts of groundwater discharge on water quality in a nearby lagoon. Re, Sacchi, Mas-Pla, Menció, and El Armani (2014) call for source control involving fertilizer use and improvements in sanitation facilities to improve water quality in the region. They also recommend extending irrigation in the area to reduce infiltration from streams, thus enhancing groundwater quality. Such insight and recommendations would not arise from the calculation of grey water footprints.

Chenoweth, Hadjikakou, and Zoumides (2014) and Launiainen et al. (2014) also find grey water footprints to be inappropriate descriptors of water pollution, as they do not account for any downstream impacts of water quality degradation. Perry (2014) notes that water quality requirements and impacts will vary with water uses. In-stream fisheries, irrigators, aquatic wildlife and hydropower plants have different water quality requirements, which are not considered in the calculation of grey water footprints. Morrison, Schulte, and Schenck (2010) consider grey water footprints to be theoretical, as they do not involve actual water volumes. In their view, summing grey water footprints with estimates of green and blue water footprints is misleading and of little value. Gawel and Bernsen (2011b) share the view that grey water footprints are fictional measures of water pollution, noting also that most wastewater is treated rather than diluted.

The optimization analysis is misguided

Schyns and Hoekstra (2014a) conduct an optimization analysis in which they endeavour to minimize Morocco's agricultural water footprint by relocating crop production across river basins. Inputs to the optimization model include the average crop areas observed from 1996 to 2005 and the authors' calculations of average, crop-specific water footprints for the same time period. The authors require the model to generate at least the same national production of each crop as in the base case, while also constraining the land available in each river basin to the base case areas. The model essentially relocates crop production activities according to calculated water footprints. There are no economic variables included in the optimization model.

The authors consider two scenarios in their optimization analysis. In scenario A, both annual and perennial crops can be relocated across river basins, while in scenario B, only annual crops can be relocated. Perennial crops in the analysis include almonds, dates, grapes, olives, oranges, sugarcane and mandarins. Annual crops include wheat, barley, maize, sugar beet and tomatoes. As expected, the optimal cropping pattern generated by the model exhibits greater shifting of crop production activities across river basins in scenario A, in which all crops are eligible to be relocated.

The optimal cropping patterns in both scenarios reflect the inappropriateness of attempting to enhance agricultural production in Morocco on the basis of water footprints. In the base case, with the exception of sugar beet and sugarcane, all 12 major crops in Morocco are produced in all eight river basins (Table 5). Such regional diversification of cropping patterns is essential in a country with limited and uncertain rainfall, and notable variability in weather patterns across and within production regions (Mrabet et al., 2012; Pala et al., 2011). In scenario A, in which all crops can be relocated, crop diversification in Morocco is severely reduced. All maize production in the country is moved to Moulouya, while barley production occurs in just three of the eight river basins (Table 6). Four of the eight river basins become one-crop basins. Farmers in Souss Massa and Tensift will produce only barley, while farmers in Sud Atlas and Loukkos will produce only wheat. Gone from those basins are the maize, fruit and tomatoes that provide crop diversity and generate income for farmers. Oum er Rbia, Moulouya and Sebou become two-crop basins in scenario A, with similar loss in diversity and reductions in farm income.

Nearly all the fruit and vegetable production in scenario A is moved to the Bouregreg basin, which will no longer produce wheat, barley or maize (Table 6). It is not clear how the farmers in Bouregreg, who produce 682,000 ha of those crops in the base case, will take on the new enterprise and manage the risk of producing only fruit and vegetables in future. Similarly, it is not clear how the many fruit and vegetable farmers in Sebou,

Table 5. Base case crop areas in Morocco representing average areas (ha, thousands), 1996–2005.

Crop	Totals	Sud Atlas	Souss Massa	Tensift	Oum er Rbia	Bouregreg	Moulouya	Sebou	Loukkos
Wheat	2874	27	42	161	909	456	140	1008	130
Barley	2208	30	175	416	738	206	283	298	60
Maize	273	9	6	60	163	20	1	13	0
Olives	512	8	25	58	161	60	34	139	29
Almonds	136	1	6	15	42	21	7	36	7
Oranges	51	1	2	5	14	3	5	17	4
Grapes	49	16	8	2	7	1	7	6	2
Dates	37	1	2	4	11	2	3	12	3
Tangerines	24	1	1	2	7	1	2	8	2
Tomatoes	22	0	1	2	7	3	1	6	1
Sugar beet	59				26	0	3	28	2
Sugarcane	15				0	1	1	10	3
Totals	6260	94	269	726	2084	775	487	1581	244

Source: Schyns and Hoekstra (2014b, appx VIII, p. 91).

Table 6. Optimal allocation of crop production in Morocco if all crops can be relocated across river basins (scenario A) (ha, thousands).

Crop	Totals	Sud Atlas	Souss Massa	Tensift	Oum er Rbia	Bouregreg	Moulouya	Sebou	Loukkos
Wheat	2787	94			807		266	1377	244
Barley	2261		269	726	1265				
Maize	221						221		
Olives	567					363		204	
Almonds	152					152			
Oranges	52					51			
Grapes	50					50			
Dates	38					38			
Tangerines	25					25			
Tomatoes	22					22			
Sugar beet	59					59			
Sugarcane	15					15			
Totals	6247	94	269	726	2071	775	487	1581	244

Source: Schyns and Hoekstra (2014b, appx VIII, p. 94).

Oum er Rbia, Tensift, and other river basins might respond to relinquishing their current activities, while producing only grains and a few olives in scenario A.

The reduction in crop diversification is less severe in scenario B, in which perennial crops cannot be relocated across river basins. However, as in scenario A, all maize production in Morocco is moved to the Moulouya basin, while barley production is concentrated in just three river basins (Table 7). Farmers in the Bouregreg basin retain 605,000 ha of wheat production in scenario B, when the relocation of perennial crops is not possible. That restriction results in a scenario that resembles the base case more closely than does scenario A. However, one still might be concerned about the proposed concentration of maize production in Moulouya and the concentration of barley production in Souss Massa, Tensift and Oum er Rbia. Wheat, barley and maize account for about 75% of the average daily intake of calories in Morocco (FAO, 2016a). Concentrating the geographical distribution of those crops could increase the likelihood of a catastrophic crop failure in future, with negative implications for household and national food security.

The optimization results depicted in scenarios A and B reflect the outcome of a selection procedure based largely on calculated water footprints, with no consideration

Table 7. Optimal allocation of crop production in Morocco if only annual crops can be relocated across river basins (scenario B) (ha, thousands).

Crop	Totals	Sud Atlas	Souss Massa	Tensift	Oum er Rbia	Bouregreg	Moulouya	Sebou	Loukkos
Wheat	2842	66			416	605	207	1353	194
Barley	2274		225	641	1409				
Maize	221						221		
Olives	512	8	25	58	161	60	34	139	29
Almonds	136	1	6	15	42	21	7	36	7
Oranges	51	1	2	5	14	3	5	17	4
Grapes	49	16	8	2	7	1	7	6	2
Dates	37	1	2	4	11	2	3	12	3
Tangerines	24	1	1	2	7	1	2	8	2
Tomatoes	22					22			
Sugar beet	59					59			
Sugarcane	15					1	1	10	3
Totals	6241	94	269	726	2065	775	487	1581	244

Source: Schyns and Hoekstra (2014b, appx VIII, p. 96).

of the issues and variables that influence both the optimal distribution of crop activities and the actual distributions that evolve over time. An optimization procedure designed to minimize an aggregate water footprint cannot generate a truly optimal allocation of crop areas. Optimality requires consideration of economics, agronomy, hydrology, the markets for inputs and outputs, and the impacts on livelihoods of any changes in existing cropping patterns. The reallocations of crop activities suggested in scenarios A and B, if implemented, would substantially disrupt agricultural production in Morocco, with potentially harmful impacts on farmers and their households in all river basins.

Schyns and Hoekstra (2014a) suggest that Morocco should consider relocating crop production, as described in scenarios A and B, to achieve the indicated water savings of 1.9 and 1.2 billion m^3 of water respectively. Those volumes might seem large at first, yet much of the estimated water savings involves rainfall that is used in place, rather than developed supplies of surface water or groundwater. In addition, half the estimated water savings in scenario A (939 of 1896 million m^3) and 80% of the estimated water savings in scenario B (939 of 1174 million m^3) are achieved by moving all maize production in Morocco to the Moulouya basin (Schyns & Hoekstra, 2014a, tab. 7). As noted above, such a concentration of maize production in a single region would increase the risk of national crop failure, while also restricting farm-level crop diversification.

One might also note that the moving of maize to Moulouya is motivated largely by the very low water footprint calculated for maize in that river basin, in comparison with the maize water footprints in other regions. In particular, the calculated water footprint for Moulouya is 1219 m^3/ton, while those for the other basins range from 3015 to 7387 m^3/ton (Table 8). The ratio of the largest to the smallest water footprint, across river basins, is 6.1 for maize, while the ratio ranges from 1.1 to 2.9 for other crops. The calculated water footprint for maize in Moulouya is based on an average crop area of just 838 ha during 1996–2005, which is less than 0.3% of the average area planted in maize in Morocco during those years (Schyns & Hoekstra, 2014b). It seems reasonable to question the validity of the very low calculated water footprint for maize in Moulouya, and to wonder whether it is appropriate to use that coefficient in an optimization analysis. Yet, lacking any information regarding the variance in crop

Table 8. Crop-specific average water footprints in Morocco river basins (m^3/ton), 1996–2005.

Crop	Ratio of high to low	Sud Atlas	Souss Massa	Tensift	Oum er Rbia	Bouregreg	Moulouya	Sebou	Loukkos
Wheat	2.9	1008	2908	3079	3168	3115	2595	3001	2329
Barley	2.8	1451	2498	3882	4054	4003	3405	3955	3043
Maize	6.1	3015	7063	5746	7387	6447	1219	6549	3178
Olives	1.1	5209	5102	5122	5063	4835	4756	4854	4651
Almonds	1.1	10,309	10,141	10,210	10,061	9618	9450	9712	9295
Oranges	1.1	545	535	538	532	499	502	509	487
Grapes	2.4	1366	1412	1420	1547	655	1002	1189	1067
Dates	1.8	4222	5917	7295	7808	4839	6640	6961	4716
Tangerines	1.1	528	517	521	515	483	486	492	471
Tomatoes	1.1	100	99	98	97	93	92	94	89
Sugar beet	1.8				124	65	106	114	101
Sugarcane	1.1					105	175	220	146

Source: Schyns and Hoekstra (2014b, appx VIII, p. 81).

yields, water use or rainfall, it is not possible to assess the validity of any of the calculated water footprints.

Throughout their analysis, Schyns and Hoekstra (2014a) provide no insight regarding the inherent variability in crop yields, water use or rainfall in Morocco, over time or across river basins. They offer their calculated water footprints without any measure of dispersion and with no consideration of statistical significance. Thus, it is not possible to know how the calculated water footprints might vary over time, or whether the water footprints in one river basin are significantly different from those in another. The single-valued estimates of crop-specific water footprints used by the authors in their optimization analysis are the products of an arithmetic exercise devoid of statistical analysis (Hoekstra et al., 2011). In a setting in which crop yields, water use and rainfall are notably stochastic, the results of such an analysis must be highly discounted.

Conclusions

Policies and investments are needed in many countries to encourage improvements in water management, increase the output of rainfed and irrigated agriculture, raise incomes in rural areas, and enhance national and household food security. Policy analysts can assist in identifying appropriate interventions by examining crop and livestock production opportunities, and developing scenarios that reflect the likely responses to changes in agricultural technology, resource availability, market opportunities, livelihood goals and climate change. Such work has guided public officials in evaluating and choosing policy options for many years in many countries.

Advising national governments on policies and investments in water resources is particularly challenging, as water is a key input in agriculture, industry and commerce, and it plays important roles in the environment and in supporting rural livelihoods. Water scarcity is increasing in many regions in which institutions are not well developed and where access to water is uneven across individuals and groups. Social and physical scientists can contribute substantially to the policy process by modelling water-use systems, analyzing alternative policies and investments, and developing recommendations based on sound science that includes the consideration of a wide range of key variables and interactions.

An assessment of water footprints is not a sufficient undertaking for advising public officials regarding policies and investments in the water sector. Water footprints contain too little information to depict the complexity of water use in any setting and they are silent on the issues of opportunity cost and scarcity value. Water footprints have no economic content and do not reflect incremental costs or benefits. They are simply ratios describing the average amount of output generated per unit of water used in a production process. In most applications, water footprints are presented with no information regarding variation and no tests of statistical significance. In such cases, an assessment of calculated water footprints is particularly inadequate for guiding policy decisions.

The water footprint assessment described in this paper does not add value to water policy considerations in Morocco. It would not be helpful to consider relocating crop production across river basins in Morocco in pursuit of a lower national water footprint in agriculture. Such a strategy would cause substantial disruption in current agricultural production patterns in the country, with potentially harmful implications for national and household food security. Agriculture is a key component of Morocco's economy and it provides employment to millions of residents who have few alternative livelihood opportunities. It would be unwise to consider relocating crop activities in the country, absent a much more thorough and comprehensive analysis. It is essential to analyze fully the economics of proposed policy alternatives, with due consideration also of agronomy, hydrology, climate change, technology, human capital, input and output markets, culture, and the likely impacts on livelihoods. A water footprint assessment is not up to the task of informing policy-makers in a manner that will enable them to reach wise decisions regarding agriculture, land and water resources, or food security.

A better way forward for Morocco would include efforts to improve crop and livestock yields, on both large and small-scale farms, and to reduce the variation in yields, over time. Investments in irrigation will be helpful in some regions, while efforts to increase productivity and reduce vulnerability in rainfed areas also are essential. Farmers use too little fertilizer and pesticides, in part because the potential financial loss in very dry years is substantial. Efforts to improve farm-level access to critical inputs, including affordable credit and crop insurance, will enhance national food security by promoting greater use of inputs that can notably increase agricultural output. Investments that enable smallholder farmers to increase average annual returns, while reducing the risk of substantial financial losses, will enhance household food security by improving the ability of households to purchase food in local markets. Complementary investments in rural infrastructure and in policies that encourage efficient water use in agriculture and other sectors will increase the likelihood of sustaining economic growth and ensuring food security for all.

Acknowledgments

This paper was given at the workshop on Virtual Water in Agricultural Products: Quantification, Limitations and Trade Policy (Lincoln, Nebraska, USA, 14-16 September 2016), which was sponsored by the OECD Co-operative Research Programme: Biological Resource Management for Sustainable Agricultural Systems whose financial support made it possible for the author to participate in the workshop.

References

Abderrazzak, B., Daoui, K., Kajji, A., Dahan, R., & Ibriz, M. (2013). Effects of supplemental irrigation and nitrogen applied on yield and yield components of bread wheat at the Saïs Region of Morocco. *American Journal of Experimental Agriculture*, *3*(4), 904–913.

African Development Bank (AfDB). (2012). *The political economy of food security in North Africa* (p. 30). AfDB Economic Brief. Abidjan, Cote d'Ivoire: The African Development Bank.

Balaghi, R., Tychon, B., Eerens, H., & Jlibene, M. (2008). Empirical regression models using NDVI, rainfall and temperature data for the early prediction of wheat grain yields in Morocco. *International Journal of Applied Earth Observations and Geoinformation*, *10*, 438–452.

Benlhabib, O., Yazar, A., Qadir, M., Louren, E., & Jacobsen, S. E. (2014). How can we improve Mediterranean cropping systems? *Journal of Agronomy and Crop Science*, *200*(5), 325–332.

Chen, W., Wu, S., Lei, Y., & Li, S. (2017). China's water footprint by province, and inter-provincial transfer of virtual water. *Ecological Indicators*, *74*, 321–333.

Chenoweth, J., Hadjikakou, M., & Zoumides, C. (2014). Quantifying the human impact on water resources: A critical review of the water footprint concept. *Hydrology and Earth System Sciences*, *18*(6), 2325–2342.

Chouchane, H., Hoekstra, A. Y., Krol, M. S., & Mekonnen, M. M. (2015). The water footprint of Tunisia from an economic perspective. *Ecological Indicators*, *52*, 311–319.

Deng, G., Ma, Y., & Li, X. (2016). Regional water footprint evaluation and trend analysis of China: Based on interregional input–output model. *Journal of Cleaner Production*, *112*, 4674–4682.

Donoso, G., Blanco, E., Franco, G., & Lira, J. (2016). Water footprints and irrigated agricultural sustainability: The case of Chile. *International Journal of Water Resources Development*, *32*(5), 738–748.

Doukkali, M. R. (2005). Water institutional reforms in Morocco. *Water Policy*, *7*(1), 71–88.

Droogers, P., Immerzeel, W. W., Terink, W., Hoogeveen, J., Bierkens, M. F. P., Van Beek, L. P. H., & Debele, B. (2012). Water resources trends in Middle East and North Africa towards 2050. *Hydrology and Earth System Sciences*, *16*(9), 3101–3114.

Duarte, R., Pinilla, V., & Serrano, A. (2014). The water footprint of the Spanish agricultural sector: 1860–2010. *Ecological Economics*, *108*, 200–207.

El Yaouti, F., El Mandour, A., Khattach, D., Benavente, J., & Kaufmann, O. (2009). Salinization processes in the unconfined aquifer of Bou-Areg (NE Morocco): A geostatistical, geochemical, and tomographic study. *Applied Geochemistry*, *24*, 16–31.

Food and Agriculture Organization of the United Nations (FAO). (2016a). *Food balance sheets for Morocco*. Retrieved December 28, 2016, from http://faostat3.fao.org/download/FB/FBS/E

Food and Agriculture Organization of the United Nations (FAO). (2016b). *International trade data for Morocco*. Retrieved December 28, 2016, from http://www.fao.org/faostat/en/#data/TP

Food and Agriculture Organization of the United Nations (FAO). (2016c). *Aquastat data sheet for Morocco*. Retrieved December 28, 2016, from http://www.fao.org/nr/water/aquastat/countries_regions/MAR/index.stm

Food and Agriculture Organization of the United Nations (FAO). (2016d). *Population data for Morocco*. Retrieved December 28, 2016, from http://www.fao.org/faostat/en/#data/OA

Food and Agriculture Organization of the United Nations (FAO). (2016e). *Crop production data for Morocco*. Retrieved December 28, 2016, from http://www.fao.org/faostat/en/#data/QC

Franke, N. A., Boyacioglu, H., & Hoekstra, A. Y. (2013). *Grey water footprint accounting: Tier 1 supporting guidelines* (Value of Water Research Report Series No. 65). Delft, The Netherlands: UNESCO-IHE.

Frija, I., Frija, A., Chebil, A., M'Hamed, H. C., Speelman, S., & Makhlouf, M. (2014). Marginal water productivity in irrigated durum wheat in semi-arid Tunisia. *Journal of Agricultural Science*, *6*(10), 84–95.

Gawel, E. (2014). Virtual water and trade: A critical economic review. In A. Bhaduri, J. Bogardi, J. Leentvaar, & S. Marx (Eds.), *The global water system in the Anthropocene* (pp. 27–43). Switzerland: Springer Water, Springer International.

Gawel, E., & Bernsen, K. (2011a). Do we really need a water footprint? Global trade, water scarcity and the limited role of virtual water. *GAIA*, *20*(3), 162–167.

Gawel, E., & Bernsen, K. (2011b). The colour of water: What does it tell us about scarcity? *GAIA*, *20*(4), 224–228.

Gawel, E., & Bernsen, K. (2013). What is wrong with virtual water trading? On the limitations of the virtual water concept. *Environment and Planning C: Government and Policy*, *31*, 168–181.

Ghanem, H. 2015. *Agriculture and rural development for inclusive growth and food security in Morocco* (Global Economy & Development Working Paper 82). Washington: The Brookings Institution, pp. 25.

Gu, Y., Xu, J., Keller, A. A., Yuan, D., Li, Y., Zhang, B., … Li, F. (2015). Calculation of water footprint of the iron and steel industry: A case study in Eastern China. *Journal of Cleaner Production*, *92*, 274–281.

Haddad, N., Duwayri, M., Oweis, T., Bishaw, Z., Rischkowsky, B., Aw Hassan, A., & Grando, S. (2011). The potential of small-scale rainfed agriculture to strengthen food security in Arab countries. *Food Security, 3*(Suppl. 1), S163–S173.

Hellegers, P., Immerzeel, W., & Droogers, P. (2013). Economic concepts to address future water supply–demand imbalances in Iran, Morocco, and Saudi Arabia. *Journal of Hydrology, 502*, 62–67.

Hoekstra, A. Y. (2014). Water for animal products: A blind spot in water policy. *Environmental Research Letters, 9*(9), 1–3.

Hoekstra, A. Y., Chapagain, A. K., Aldaya, M. M., & Mekonnen, M. M. (2009). *Water footprint manual: State of the Art 2009*. Enschede, The Netherlands: Water Footprint Network.

Hoekstra, A. Y., Chapagain, A. K., Aldaya, M. M., & Mekonnen, M. M. (2011). *The water footprint assessment manual: Setting the global standard*. London: Earthscan.

Hoekstra, A. Y., Chapagain, A. K., & Zhang, G. (2016). Water footprints and sustainable water allocation. *Sustainability (Switzerland), 8*(1), 1–6.

Huang, J., Xu, C., Ridoutt, B. G., & Chen, F. (2015). Reducing agricultural water footprints at the farm scale: A case study in the Beijing region. *Water (Switzerland), 7*(12), 7066–7077.

Jacobsen, S.-E., Jensen, C. R., & Liu, F. (2012). Improving crop production in the arid Mediterranean climate. *Field Crops Research, 128*, 34–47.

Jobbins, G., Kalpakian, J., Chriyaa, A., Legrouri, A., & El Mzouri, E. H. (2015). To what end? Drip irrigation and the water–energy–food nexus in Morocco. *International Journal of Water Resources Development, 31*(3), 393–406.

Karrou, M., & Oweis, T. (2014). Assessment of the severity and impact of drought spells on rainfed cereals in Morocco. *African Journal of Agricultural Research, 9*(49), 3519–3530.

Konar, M., Reimer, J. J., Hussein, Z., & Hanasaki, N. (2016). The water footprint of staple crop trade under climate and policy scenarios. *Environmental Research Letters, 11*(3), 1–15.

Launiainen, S., Futter, M. N., Ellison, D., Clarke, N., Finér, L., Högbom, L., . . . Ring, E. (2014). Is the water footprint an appropriate tool for forestry and forest products: The Fennoscandian case. *Ambio, 43*(2), 244–256.

Lovarelli, D., Bacenetti, J., & Fiala, M. (2016). Water footprint of crop productions: A review. *Science of the Total Environment, 548–549*, 236–251.

Lyamani, A., Silué, D., & Guéi, R. G. (2011). Morocco: The visible hand. In P. Van Mele, J. W. Bentley, & R. G. Guéi (Eds.), *African seed enterprises: Sowing the seeds of food security* (Chapter 8, pp. 133–141). Rome: Food and Agriculture Organization of the United Nations, and Cotonou: Africa Rice Center.

Magnan, N., Lybbert, T. J., Mrabet, R., & Fadlaoui, A. (2011). The quasi-option value of delayed input use under catastrophic drought risk: The case of no-till in Morocco. *American Journal of Agricultural Economics, 93*(2), 498–504.

Mahe, G., Lienou, G., Descroix, L., Bamba, F., Paturel, J. E., Laraque, A., . . . Khomsi, K. (2013). The rivers of Africa: Witness of climate change and human impact on the environment. *Hydrological Processes, 27*(15), 2105–2114.

Mekonnen, M. M., & Hoekstra, A. Y. (2014a). Water footprint benchmarks for crop production: A first global assessment. *Ecological Indicators, 46*, 214–223.

Mekonnen, M. M., & Hoekstra, A. Y. (2014b). Water conservation through trade: The case of Kenya. *Water International, 39*(4), 451–468.

Mohamed, B. (2013). *Better economics: Supporting climate change adaptation with stakeholder analysis: A case study of Morocco* (pp. 28). London: International Institute for Environment and Development.

Morrison, J., Schulte, P., & Schenck, R. (2010). *Corporate water accounting: An analysis of methods and tools for measuring water use and its impact*. Oakland, CA: United Nations Environment Programme and Pacific Institute.

Mrabet, R., Moussadek, R., Fadlaoui, A., & Van Ranst, E. (2012). Conservation agriculture in dry areas of Morocco. *Field Crops Research, 32*, 84–94.

Murphy, E., de Boer, I. J. M., van Middelaar, C. E., Holden, N. M., Shalloo, L., Curran, T. P., & Upton, J. (2017). Water footprinting of dairy farming in Ireland. *Journal of Cleaner Production, 140*, 547–555.

Pahlow, M., Snowball, J., & Fraser, G. (2015). Water footprint assessment to inform water management and policy making in South Africa. *Water SA, 41*(3), 300–313.

Pala, M., Oweis, T., Benli, B., De Pauw, E., El Mourid, M., Karrou, M., ... Zencirci, N. (2011). *Assessment of wheat yield gap in the Mediterranean: Case studies from Morocco, Syria, and Turkey* (pp. 36). Aleppo, Syria: International Center for Agricultural Research in the Dry Areas (ICARDA).

Pellicer-Martínez, F., & Martínez-Paz, J. M. (2016). The water footprint as an indicator of environmental sustainability in water use at the river basin level. *Science of the Total Environment, 571*, 561–574.

Perry, C. (2014). Water footprints: Path to enlightenment, or false trail? *Agricultural Water Management, 134*, 119–125.

Re, V., Sacchi, E., Martin-Bordes, J. L., Aureli, A., El Hamouti, N., Bouchnan, R., & Zuppi, G. M. (2013). Processes affecting groundwater quality in arid zones: The case of the Bou-Areg coastal aquifer (North Morocco). *Applied Geochemistry, 34*, 181–198.

Re, V., Sacchi, E., Mas-Pla, J., Menció, A., & El Armani, N. (2014). Identifying the effects of human pressure on groundwater quality to support water management strategies in coastal regions: A multi-tracer and statistical approach (Bou-Areg region, Morocco). *Science of the Total Environment, 500–501*, 211–223.

Rudenko, I., Bekchanov, M., Djanibekov, U., & Lamers, J.P.A. (2013). The added value of a water footprint approach: Micro- and macroeconomic analysis of cotton production, processing and export in water bound Uzbekistan. *Global and Planetary Change, 110*, 143–151.

Ryan, J., Ibrikci, H., Delgado, A., Torrent, J., Sommer, R., & Rashid, A. (2012). Significance of phosphorus for agriculture and the environment in the West Asia and North Africa region. *Advances in Agronomy, 114*, 91–153.

Ryan, J., Ibrikci, H., Sommer, R., & McNeill, A. (2009). Nitrogen in rainfed and irrigated cropping systems in the Mediterranean region. *Advances in Agronomy, 104*, 53–136.

Saadi, S., Todorovic, M., Tanasijevic, L., Pereira, L. S., Pizzigalli, C., & Lionello, P. (2015). Climate change and Mediterranean agriculture: Impacts on winter wheat and tomato crop evapotranspiration, irrigation requirements and yield. *Agricultural Water Management, 147*, 103–115.

Schilling, J., Frier, K. P., Hertig, E., & Scheffran, J. (2012). Climate change, vulnerability and adaptation in North Africa with a focus on Morocco. *Agriculture, Ecosystems and Environment, 156*, 12–26.

Schyns, J. F., Hamaideh, A., Hoekstra, A. Y., Mekonnen, M. M., & Schyns, M. (2015). Mitigating the risk of extreme water scarcity and dependency: The case of Jordan. *Water (Switzerland), 7* (10), 5705–5730.

Schyns, J. F., & Hoekstra, A. Y. (2014a). The added value of water footprint assessment for national water policy: A case study of Morocco. *PLoS One, 9*(6), e99705.

Schyns, J. F., & Hoekstra, A. Y. (2014b). *The water footprint in Morocco: The added value of Water Footprint Assessment for national water policy* (Value of Water Research Report Series No. 67). Delft, The Netherlands: UNESCO-IHE.

Shi, J., Liu, J., & Pinter, L. (2014). Recent evolution of China's virtual water trade: Analysis of selected crops and considerations for policy. *Hydrology and Earth System Sciences, 18*(4), 1349–1357.

Snoussi, M., Haïda, S., & Imassi, S. (2002). Effects of the construction of dams on the water and sediment fluxes of the Moulouya and the Sebou Rivers, Morocco. *Regional Environmental Change, 3*, 3–12.

Tanasijevic, L., Todorovic, M., Pereira, L. S., Pizzigalli, C., & Lionello, P. (2014). Impacts of climate change on olive crop evapotranspiration and irrigation requirements in the Mediterranean region. *Agricultural Water Management, 144*, 54–68.

Tekken, V., & Kropp, J. P. (2012). Climate-driven or human-induced: Indicating severe water scarcity in the Moulouya River Basin (Morocco). *Water (Switzerland)*, *4*(4), 959–982.

Terink, W., Immerzeel, W. W., & Droogers, P. (2013). Climate change projections of precipitation and reference evapotranspiration for the Middle East and Northern Africa until 2050. *International Journal of Climatology*, *33*(14), 3055–3072.

Vanham, D., Gawlik, B. M., & Bidoglio, G. (2017). Food consumption and related water resources in Nordic cities. *Ecological Indicators*, *74*, 119–129.

Vanham, D., Mak, T. N., & Gawlik, B. M. (2016). Urban food consumption and associated water resources: The example of Dutch cities. *Science of the Total Environment*, *565*, 232–239.

Wichelns, D. (2004). The policy relevance of virtual water can be enhanced by considering comparative advantages. *Agricultural Water Management*, *66*(1), 49–63.

Wichelns, D. (2010a). Virtual water: A helpful perspective, but not a sufficient policy criterion. *Water Resources Management*, *24*(10), 2203–2219.

Wichelns, D. (2010b). Virtual water and water footprints offer limited insight regarding important policy questions. *International Journal of Water Resources Development*, *26*(4), 639–651.

Wichelns, D. (2011a). Do the virtual water and water footprint perspectives enhance policy discussions? *International Journal of Water Resources Development*, *27*(4), 625–637.

Wichelns, D. (2011b). Virtual water and water footprints: Compelling notions, but notably flawed. *GAIA*, *20*(3), 171–175.

Wichelns, D. (2015a). Water productivity and food security: Considering more carefully the farm-level perspective. *Food Security*, *7*(2), 247–260.

Wichelns, D. (2015b). Virtual water and water footprints do not provide helpful insight regarding international trade or water scarcity. *Ecological Indicators*, *52*, 277–283.

Wichelns, D. (2015c). Virtual water and water footprints: Overreaching into the discourse on sustainability, efficiency, and equity. *Water Alternatives*, *8*(3), 396–414.

World Bank. (2015). *World DataBank*. Retrieved December 28, 2016, from http://databank.worldbank.org/ddp/home.do

Ziolkowska, J. R. (2015). Shadow price of water for irrigation: A case of the High Plains. *Agricultural Water Management*, *153*, 20–31.

Zoumides, C., Bruggeman, A., Hadjikakou, M., & Zachariadis, T. (2014). Policy-relevant indicators for semi-arid nations: The water footprint of crop production and supply utilization of Cyprus. *Ecological Indicators*, *43*, 205–214.

Future crop yields and water productivity changes for Nebraska rainfed and irrigated crops

Yaqiong Lu, Xianyu Yang and Lara Kueppers

ABSTRACT

We assessed future rainfed and irrigated crop yield and water productivity changes in Nebraska across multiple climate and emission scenarios using an empirical modeling approach. We found rainfed crops showed slightly increased crop water productivity while irrigated crops showed no change or decreased water productivity. Contrary to U.S.-wide studies reporting declines in crop yields, we projected Nebraska crop yields to increase overall with greatest increases in current rainfed fields due to combined effects from maximum and minimum temperatures. However, the increased rainfed yields are not sufficient to fully close the gap between rainfed and irrigated yields.

Abbreviations: USDA: U.S. Department of Agriculture; RegCM4.3: ICTP Regional Climate Model version 4.3; NCEP: National Centers for Environmental prediction; DOE: U.S. Department of Energy; CGCM: Canadian Climate Centre general circulation model; GFDL: Geophysical Fluid Dynamics Laboratory general circulation model; CRCM: Canadian Climate Centre regional climate model; CCSM: National Center for Atmospheric Research general circulation model; HRM3: Hadley Centre's Regional Model 3; HADCM3: Hadley Centre's general circulation model; WRFG: the NCAR Weather Research and Forecasting model; CCCma: Canadian Centre for Climate Modelling and Analysis; CanESM2: Canadian Centre Earth System Model 2; ICHEC-EC: A European community Earth-System Model; IPCC: Intergovernmental Panel on Climate Change; RMSE: Root Mean Square Error

Introduction

How crop yields will change under future climate is a fundamental question for regional and global food security (Lesk, Rowhani, & Ramankutty, 2016; Lobell et al., 2008; Rosenzweig & Parry, 1994). Warmer temperatures and unevenly distributed precipitation could result in large local changes in crop yields. While longer, warmer growing seasons could benefit some crops and regions, crop yields decline when temperature exceeds certain thresholds, and therefore some studies project severe damage to U.S. crop yields under climate change (Lobell & Asner, 2003; Schlenker & Roberts, 2009). Besides reductions in mean yields, one study found that one third of crop yield variability is explained by temperature and precipitation variability (Ray, Gerber, MacDonald, & West, 2015), indicating that variation in yields is at least partially tied to climate variation. Our prior work suggests that future warming and climate variability would increase variability in corn and soy yields in the U.S. Corn Belt (Thompson et al., 2017).

At the same time, irrigation water demand by crops is expected to increase with warmer temperatures. Two-thirds of global irrigated cropland will possibly suffer from greater irrigation requirements, and Nebraska was predicted to have a 5–15% increase in irrigation requirement with anthropogenic climate change (Doll, 2002). However, ground water depletion is leading to a supply and demand imbalance (Famiglietti & Rodell, 2013). In the U.S. Southern High Plains, 35% of the area will be unable to support irrigation within the next 30 years under current ground water depletion rates (Scanlon et al., 2012). Water limitations have been projected to return 20–60 M ha of irrigated cropland to rainfed cropland globally under representative concentration pathway 8.5 (RCP85), resulting in a further 43% – 112% loss of food production below that due to temperature and precipitation changes only (Elliott et al., 2014).

Nebraska is one of the largest agricultural states in the United States. Ninety-three per cent of Nebraska's land area is farms and ranches. Nebraska also has the largest area of irrigated cropland in the United States (3,357,903 ha. in 2013; USDA NASS) thanks to the Ogallala aquifer that extends underneath the state. Farm marketing contributed over $21 billion to Nebraska's economy in 2011 and accounted for 5.8% of the U.S. total farm income (Nebraska Department of Agriculture, 2014). Agriculture in Nebraska also contributes substantially to international markets and the bioenergy industry. For example, 50% of Nebraska's wheat is exported to international markets annually. In 2012, Nebraska ranked second in ethanol production capacity, and over 40% of the state's corn crop was utilized in ethanol production (Nebraska Department of Agriculture, 2014).

Compared to the numerous studies on crop yield changes at continental scales (Lobell et al., 2013, 2014; Naylor, Falcon, Rochberg, & Wada, 2001; You, Rosegrant, Wood, & Sun, 2009), there has been less focus on how climate changes could affect individual localities such as Nebraska's irrigated and rainfed crop yields. Mendelsohn, Nordhaus, and Shaw (1994) and Reilly et al. (2003) both found that future climate change would be harmful to yields in the southern United States due to increasing temperature, and that western arid regions would suffer more than central and eastern Nebraska (Mendelsohn et al., 1994). To facilitate Nebraska water management decisions, it is also important to understand how crop water productivity (kilograms of crop per cubic meter of water input) (Amarasinghe & Smakhtin, 2014) changes with regional climate change, for both rainfed and irrigated crops.

In this paper, we report how projected regional temperature and precipitation changes affect corn, soybean, and winter wheat yields, as well as crop water productivity (water productivity) in Nebraska, distinguishing rainfed and irrigated crops. We focused on these three crops because they are among the top ten commodities in 2011 state cash receipts. Further, temperature and precipitation account for more than 60% of the variation in yields of these crops in our statistical models, indicating that they are sensitive to climate changes. Nebraska has a dry to wet precipitation gradient, extending from west to east, therefore

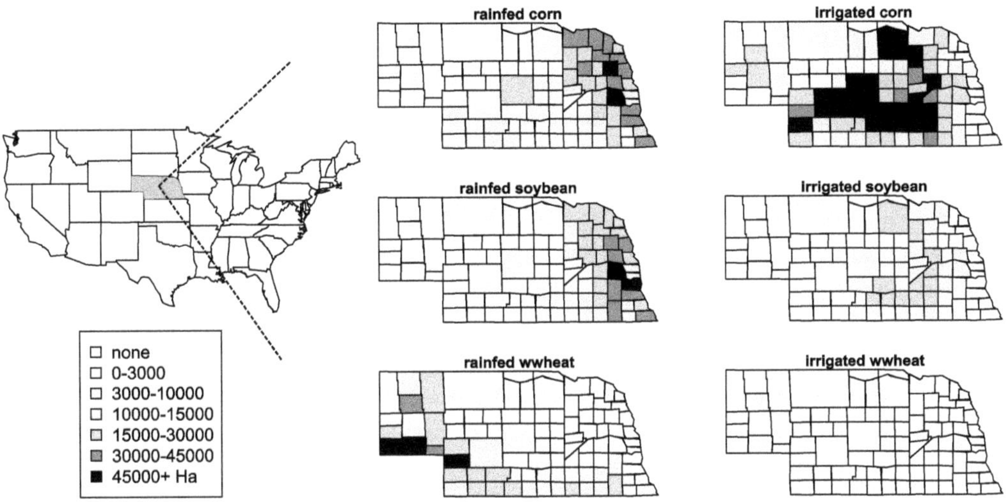

Figure 1. 1982–2012 averaged harvested Hectares of rainfed and irrigated corn, soybean, and winter wheat in Nebraska at the county level (Source data: USDA NASS).

rainfed corn and soybeans are mainly grown in the eastern part of Nebraska (Figure 1). State wide, irrigated corn accounts for 64% of corn yield, while irrigated soybeans account for 38% of soybean yield. Winter wheat is a dryland crop planted mainly in western Nebraska; irrigated winter wheat accounts for only 6% of total yield. We used 31 years of yield (1982–2012) and climate data to develop step-wise, second-order polynomial crop models at the state level for rainfed and irrigated corn, soybean, and winter wheat, and applied them to multiple scenarios of future climate change in Nebraska.

Methods

We took an empirical modeling approach, in which historical crop yield and climate data are used to train a set of statistical crop models, and then the statistical models are used with regional climate change projections to forecast crop yields for the middle of the 21st century. We parameterized the statistical crop models using training data for the years 1982 to 2012. The yield training data are the USDA National Agricultural Statistics Service (NASS) historical crop yields at the county-level (www.nass.usda.gov).[1] We generated the climate training data by running a regional climate model (RegCM4.3; Giorgi et al., 2012) with boundary condition forcing from global reanalysis data (NCEP/DOE Reanalysis 2 product; Kanamitsu et al., 2002). The time series anomaly for the RegCM4.3/NCEP simulation showed no significant difference with observation-based data.

To quantify the future change in yields, we used historical and future RCM output to drive the parameterized crop models. These RCM outputs were obtained from the North American Regional Climate Change Assessment Program database (NARCCAP) and

Coordinated Regional Climate Downscaling Experiment (CORDEX). The NARCCAP (Mearns et al., 2012; Mearns, Gutowski, & Jones, 2007; Mearns et al., 2009) simulations use the IPCC A2 scenarios, and CORDEX (Martynov et al., 2013) uses multiple IPCC Representative Concentration Pathways (RCPs). The NARCCAP simulations were driven by boundary conditions from six alternate global climate model simulations (RegCM with forcing from CGCM and GFDL, CRCM with forcing from CCSM and CGCM3, HRM3 with forcing from HADCM3, WRFG with forcing from CCSM). The CORDEX models used in our study are CCCma-CanESM2 and ICHEC-EC with RCP4.5 and RCP8.5. The simulation years included historical (1968–1998) and mid-century (2038–2069) periods. To correct for differences between the training and hindcast/forecast climate data, we applied a quantile-based bias correction (see details in Thompson et al., 2017).

We developed stepwise, second-order polynomial regression models that also considered interactions between climate variables, while avoiding over-fitting. By including the interaction terms, we expected to better account for extreme events not captured by the second-order terms. Each yield model was composed of three predictor variables with one month for monthly maximum temperature (T_{max}), one month for monthly minimum temperature (T_{min}), and one month for monthly precipitation (Precip). The full equation is given in Equation (1).

$$Yield = a_0 + a_1 year + a_2 year^2 + a_3 T_{max} + a_4 T_{max}^2 + a_5 T_{min} + a_6 T_{min}^2 \\ + a_7 Precip + a_8 Precip^2 + a_9 T_{max} Precip + a_{10} T_{min} Precip + a_{11} T_{max} T_{min} \quad (1)$$

Because using the full Equation (1) could result in over-fitting due to the limited number of years (31) used to construct the model, we used the stepwise function in R to automatically build models by adding or removing terms. For each crop, we generated a range of regression models by using permutations of temperature and precipitation from different months. Each regression model uses the best combination of terms based on Akaike information criterion (AIC) selection for a given month of tmax, tmin, and precipitation. Then we selected the best five regression models based on a historical out-of-sample error analysis and these criteria: 1) Any two monthly climate predictors should not be highly correlated over time ($r < 0.3$), 2) no autocorrelation in errors, 3) no heteroscedasticity in errors, 4) RMSE reduced compared to the baseline model (which included 'year' terms only), 5) no negative future yield projections, and 6) $R^2 > 0.6$ for an individual climate predictor and crop yield. We used the five crop regression models to represent different stages important to crop growth rather than using one single regression model with one set of monthly predictors.

We set up and selected statistical models for irrigated crops separately, but excluded precipitation terms from the models. The full model for irrigated crops is:

$$Yield = a_0 + a_1 year + a_2 year^2 + a_3 T_{max} + a_4 T_{max}^2 + a_5 T_{min} + a_6 T_{min}^2 \\ + a_7 T_{max} T_{min} \quad (2)$$

Water productivity measures the efficiency of converting water input into crop yield. Water productivity was calculated as kilograms of crop per cubic meter of water input. The water input for rainfed crops is the growing season (April to September for corn and soybean, October to June for winter wheat) total precipitation. While for irrigated crops, the water input is the growing season total precipitation plus the total irrigation. Using the actual irrigation amount in each year would be ideal but was not possible because these observational data are not available. Instead we used data from a Nebraska irrigation survey in 2013 (USDA NASS), where total growing season irrigation water used for corn, soybean, and winter wheat were 3050 m^3/ha, 2740 m^3/ha, and 2130 m^3/ha respectively. We used the same irrigation rate to calculate both the historical and future water productivity, meaning that interannual variability in historical and future irrigation rate are missing from our water productivity analysis, and change in water productivity is due only to changes in yield and precipitation.

Results

Across multiple future emissions scenarios, a diversity of climate model representations of climate change, and multiple yield model permutations, we found that irrigated crop yields and water productivity are generally less sensitive to climate changes than rainfed crop yields in Nebraska (Figure 2). Averaged across three emission scenarios, the rainfed crop mean yield increased by 3.0%, 11.5%, and 8.7% for corn, soybean, and winter wheat respectively, while irrigated crop yield mean increased by 1.2% and 3.6% for corn and winter wheat, and decreased by 0.1% for soybean. Rainfed crops had an even greater increase in interannual yield variability. The rainfed crop yield standard deviation increased by 11.4%, 31.4%, and 67.0% for corn, soybean, and winter wheat respectively, and while irrigated yield standard deviations increased 51.5% for winter wheat, corn and soybean yield standard deviations changed little (–0.1% and 1.3%).

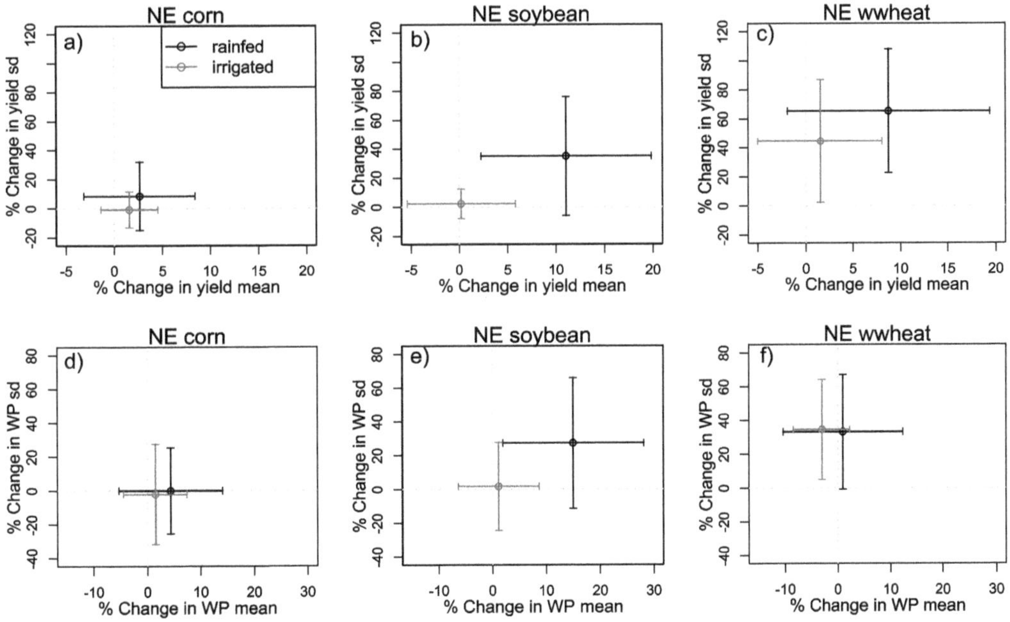

Figure 2. Percentage change in standard deviation (sd) and in mean (a) corn yield, (b) soybean yield, (c) winter wheat yield, (d) corn water productivity, (e) soybean water productivity, and (f) winter wheat water productivity for irrigated (blue) and rainfed (black) fields in Nebraska (NE). The error bars show one standard deviation of the mean across five crop yield models and 10 climate models for three emissions scenarios (6 climate models with A2, 2 with RCP45, and 2 with RCP85).

Due to the small changes in irrigated corn and soybean yields, their water productivity did not change much (1.5% and 1.2% for mean water productivity, −2.2% and 1.7% for its standard deviation). Rainfed corn and soybeans showed a greater increase (or less decrease) in mean and standard deviation of water productivity (4.4% and 15% for mean water productivity, −0.1% and 27.3% for water productivity standard deviation). For winter wheat, the water productivity standard deviation increased similarly for rainfed (33%) and irrigated (34%) fields. The increase in winter wheat mean yield did not result in a large increase in mean water productivity, with the irrigated winter wheat mean water productivity decreasing by 3.1%, and rainfed winter wheat mean water productivity increasing by only 1%.

Even though the rainfed crops had greater yield increases than irrigated crops, rainfed crop yields were still projected to be less than irrigated crop yields (Figure 3). Future irrigated corn, soybean, and winter wheat yields were 10.7 T/ha, 3.5 T/ha, 3.8 T/ha, and rainfed yields were 9.2 T/ha, 3 T/ha, 2.8 T/ha averaged across all scenarios. In fact, future rainfed crop yields did not exceed historical irrigated crops yields. At the same time, future rainfed crops had greater water productivity than future irrigated crops, except for winter wheat. Future rainfed corn, soybean, and winter wheat water

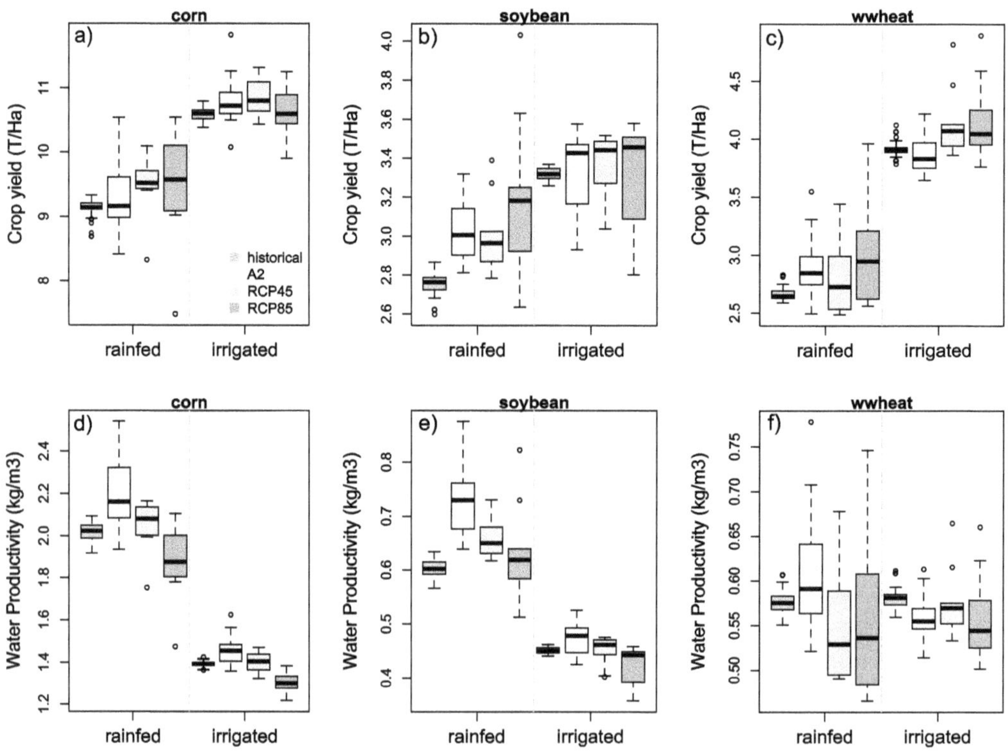

Figure 3. Boxplots for rainfed and irrigated (a) corn yield, (b) soybean yield, (c) winter wheat yield, (d) corn water productivity, (e) soybean water productivity, and (f) winter wheat water productivity under historical and future climate scenarios. Each box represents historical or future annual averaged values across five crop models and six climate models (for historical and A2) or two climate models (for RCP45 and RCP85).

productivity were 2.1 kg/m^3, 0.7 kg/m^3, 0.6 kg/m^3, and irrigated water productivities were 1.4 kg/m^3, 0.5 kg/m^3, 0.6 kg/m^3 averaged across all scenarios.

The crop yield and water productivity projections differed among the three emission scenarios. RCP85 showed the highest future rainfed crop yields (9.5 T/ha, 3.2 T/ha, 3.0 T/ha) and also resulted low water productivity (1.87 kg/m^3, 0.63 kg/m^3, 0.56 kg/m^3) for corn, soybean, and winter wheat. For irrigated crops, RCP85 did not result in the highest crop yield, and generated the lowest water productivity for corn (1.3 kg/m^3), soybean (0.42 kg/m^3), and winter wheat (0.56 kg/m^3). A2 showed a slightly greater increase in rainfed crop yields and had the highest water productivity. Irrigated corn and soybean also showed the highest water productivity under the A2 emission scenarios.

To understand the individual and combined effects of the climate variables on yield changes, we applied the future values of the individual and combined temperature and precipitation variables one at a time, while keeping other variables at the historical values. For rainfed crops, we found T_{max} and T_{min} together generated yields that most closely matched the full projections (Figure 4). Mean rainfed yield changes projected using only T_{max} and T_{min} were highly correlated with mean yield changes projected using all climate factors for corn (0.84), soybean (0.95), and winter wheat (0.91). T_{max} and T_{min} together also predicted yield standard deviation changes well; the correlation with the full projected yield standard deviation changes were 0.59 for corn, 0.97 for soybean, and 0.72 for winter wheat, suggesting that changes in T_{max} and T_{min} not only determined the magnitude of mean yield changes, but also determined a large fraction

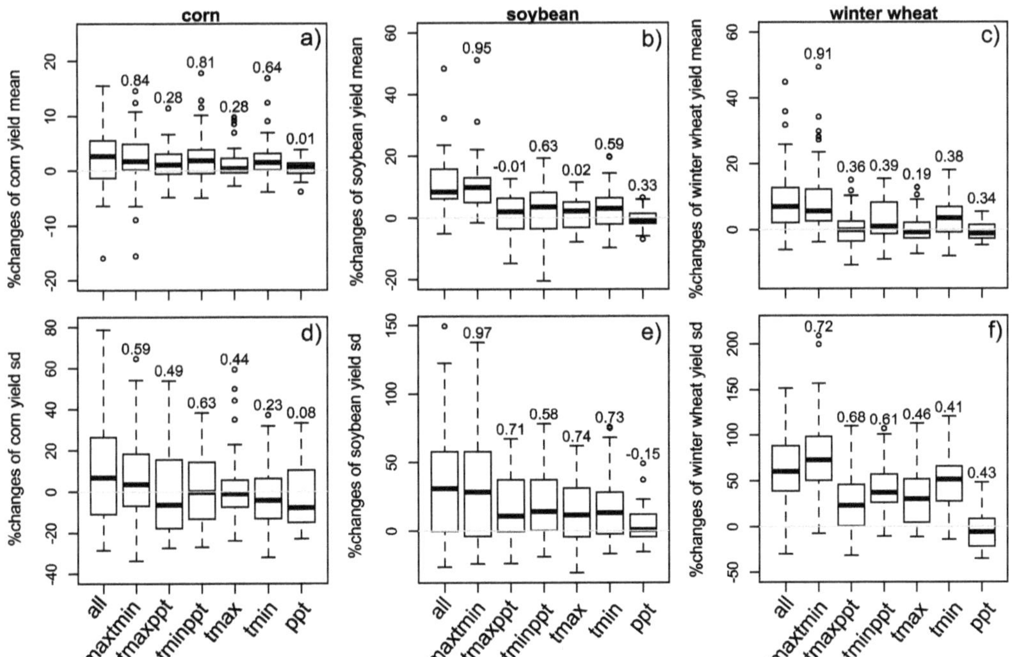

Figure 4. Rainfed crop mean yield and standard deviation changes based on changes in individual or combined climate variables. The values at the top of each box give the correlation between yield changes in the partial models and those using changes in all climate variables.

of the interannual variation. For irrigated crops, T_{min} or T_{max} alone did not show strong correlation with the full projection, where the full projection used the combined effects of T_{max} and T_{min} (Equation (2)).

Discussion

Nebraska has the largest acreage of irrigated cropland in the United States (3,357,903 ha. in 2013; USDA NASS). Understanding how irrigated crop yields change in response to a range of climate scenarios is essential to anticipating changes in food security and the agricultural economy of the state. Using an empirical modeling approach, we showed that irrigated crops are less sensitive to climate changes than rainfed crops. On average, across emission scenarios and climate models, rainfed crops showed a greater increase in both mean yield and interannual variability in yield, as did crop water productivity. The crop yield and water productivity projections differed among the three emission scenarios. However, all emission scenarios produce a wider range of future crop yield and water productivity values than the historical period, suggesting increasing uncertainty in future yield projections. When looking at the averaged results, RCP85 showed the highest future rainfed crop yields but also low water productivity, while A2 showed a slight increase in rainfed crop yields but had the highest water productivity.

Despite the lower sensitivity to climate change, irrigated crops still showed much higher crop yields than rainfed crops. The high irrigated yield coupled with the large water inputs in irrigated cropland often resulted low water productivity. Among the three crops, we found that irrigated winter wheat is the only crop with water productivity similar to its rainfed counterpart. However, the current irrigated winter wheat area is very small compared with that of irrigated corn and soybean. Our results suggest a benefit of conserving water while maintaining high yields if the irrigated winter wheat area were expanded.

Many studies have shown that future climate could reduce the United States crop yield due to the warmer climate and an earlier but warmer growing season. The magnitude of the decrease varies with emission scenario and the future period of focus. For example, averaged corn, soybean, and cotton yield in US have been projected to decrease by 30–46% before the end of the 21st century under the lowest (B1) warming scenario, with severe decreases (63–82%) under the rapid warming scenario (A1FI) (Schlenker & Roberts, 2009). Declines have been expected to be smaller mid-century (2030–2050), when US corn yields are projected to decrease by 18% (Urban, Roberts, Schlenker, & Lobell, 2012). We selected five crop models with the best estimation of the yield and used seven climate models and three emission scenarios to capture large uncertainties in crop yield projections. In our projections, we found corn, soybean, and winter wheat yields in Nebraska could actually increase on average, especially for the rainfed crops. Differences between state-level and national-level yield change projections suggest large regional variations in yield sensitivity within the US and highlight a need for sate-level decision making to consider regionally specific yield projections.

With a set of sensitivity analyses, we determined that T_{max} and T_{min} together played the most important role in increasing rainfed crop yield. One reason might be that the current locations where rainfed corn and soybean are grown are fairly wet regions in eastern Nebraska, where summer precipitation is not historically limiting. These regions averaged

800 – 900 mm annual precipitation (1982–2012), and future precipitation projections did not vary much from the historical means (< 1 mm/month). When the statistical models were trained with the high historical precipitation, the role of small future precipitation variations was not as important as the larger temperature variations (up to 6°C/month).

Conclusions

In summary, our statistical crop models for corn, soybean, and winter wheat in Nebraska, revealed that irrigated crops are less sensitive to climate changes than rainfed crops, benefitting less from warmer temperatures, but also suffering less from increased interannual variability in yield. For the rainfed crops, combined changes in T_{max} and T_{min} drove the increased mean yield and variability. Therefore, while increased rainfed yields are not sufficient to fully close the gap between rainfed and irrigated yields, maintaining – rather than expanding – the current level of irrigation in Nebraska could allow an increase state-wide crop production thanks to increases in rainfed yields.

Notes

1. Abbreviations used in the text:
   USDA: U.S. Department of Agriculture
   RegCM4.3: ICTP Regional Climate Model version 4.3
   NCEP: National Centers for Environmental prediction
   DOE: U.S. Department of Energy
   CGCM: Canadian Climate Centre general circulation model
   GFDL: Geophysical Fluid Dynamics Laboratory general circulation model
   CRCM: Canadian Climate Centre regional climate model
   CCSM: National Center for Atmospheric Research general circulation model
   HRM3: Hadley Centre's Regional Model 3
   HADCM3: Hadley Centre's general circulation model
   WRFG: the NCAR Weather Research and Forecasting model
   CCCma: Canadian Centre for Climate Modelling and Analysis
   CanESM2: Canadian Centre Earth System Model 2
   ICHEC-EC: A European community Earth-System Model
   IPCC: Intergovernmental Panel on Climate Change
   RMSE: Root Mean Square Error

Acknowledgments

We acknowledge the World Climate Research Programme's Working Group on Regional Climate, and the Working Group on Coupled Modelling, former coordinating body of CORDEX and responsible panel for CMIP5. We thank the individual climate modeling centers producing and making available their model output (CCCma-CanESM2 and ICHEC-EC). We also acknowledge North American Regional Climate Change Assessment Program (NARCCAP) for providing climate model output. NARCCAP is funded by the National Science Foundation (NSF), the U.S. Department of Energy (DoE), the National Oceanic and Atmospheric Administration (NOAA), and the U.S. Environmental Protection Agency Office of Research and Development (EPA). This work was supported by USDA AFRI under grant number 2012-68002-19872. This paper was given at the workshop on Virtual Water in Agricultural Products: Quantification, Limitations and Trade Policy (Lincoln, Nebraska, USA, 14-16 September 2016), which was sponsored by the OECD Co-operative Research Programme:

Biological Resource Management for Sustainable Agricultural Systems whose financial support made it possible for Lara Kueppers to participate in the workshop.

Disclosure statement

No potential conflict of interest was reported by the authors.

Funding

This work was supported by the USDA AFRI; [2012-68002-19872].

References

Amarasinghe, U. A., & Smakhtin, V. (2014). Water productivity and water footprint: Misguided concepts or useful tools in water management and policy? *Water International*, *39*(7), 1000–1017.

Doll, P. (2002). Impact of climate change and variability on irrigation requirements: A global perspective. *Climatic Change*, *54*(3), 269–293.

Elliott, J., Deryng, D., Mueller, C., Frieler, K., Konzmann, M., Gerten, D., … Wisser, D. (2014). Constraints and potentials of future irrigation water availability on agricultural production under climate change. *Proceedings of the National Academy of Sciences of the United States of America*, *111*(9), 3239–3244.

Famiglietti, J. S., & Rodell, M. (2013). Water in the balance. *Science*, *340*(6138), 1300–1301.

Giorgi, F., Coppola, E., Solmon, F., Mariotti, L., Sylla, M. B., Bi, X., … Brankovic, C. (2012). RegCM4: Model description and preliminary tests over multiple CORDEX domains. *Climate Research*, *52*(1), 7–29.

Kanamitsu, M., Ebisuzaki, W., Woollen, J., Yang, S. K., Hnilo, J. J., Fiorino, M., & Potter, G. L. (2002). NCEP-DOE AMIP-II Reanalysis (R-2). *Bulletin of the American Meteorological Society*, *83*(11), 1631–1643.

Lesk, C., Rowhani, P., & Ramankutty, N. (2016). Influence of extreme weather disasters on global crop production. *Nature*, *529*(7584), 84–87.

Lobell, D. B., & Asner, G. P. (2003). Climate and management contributions to recent trends in US agricultural yields. *Science*, *299*(5609), 1032.

Lobell, D. B., Burke, M. B., Tebaldi, C., Mastrandrea, M. D., Falcon, W. P., & Naylor, R. L. (2008). Prioritizing climate change adaptation needs for food security in 2030. *Science*, *319*(5863), 607–610.

Lobell, D. B., Hammer, G. L., McLean, G., Messina, C., Roberts, M. J., & Schlenker, W. (2013). The critical role of extreme heat for maize production in the United States. *Nature Climate Change*, *3*(5), 497–501.

Lobell, D. B., Roberts, M. J., Schlenker, W., Braun, N., Little, B. B., Rejesus, R. M., & Hammer, G. L. (2014). Greater sensitivity to drought accompanies maize yield increase in the US midwest. *Science*, *344*(6183), 516–519.

Martynov, A., Laprise, R., Sushama, L., Winger, K., Separovic, L., & Dugas, B. (2013). Reanalysis-driven climate simulation over CORDEX North America domain using the Canadian regional climate model, version 5: Model performance evaluation. *Climate Dynamics*, *41*(11–12), 2973–3005.

Mearns, L. O., Arritt, R., Biner, S., Bukovsky, M. S., McGinnis, S., Sain, S., … Snyder, M. (2012). The North American regional climate change assessment program overview of phase I results. *Bulletin of the American Meteorological Society*, *93*(9), 1337–1362.

Mearns, L. O., Gutowski, W., Jones, R., Leung, L.-Y., McGinnis, S., Nunes, A. M. B., & Qian, Y. (2009). A regional climate change assessment program for North America. *Eos, Transactions American Geophysical Union*, *90*(36), 311.

Mearns, L. O., Gutowski, W. J., & Jones, R. (2007). *The North American regional climate change assessment program dataset*. Boulder, CO: National Center for Atmospheric Research Earth System Grid data portal.

Mendelsohn, R., Nordhaus, W. D., & Shaw, D. (1994). The impact of global warming on agriculture: A ricardian analysis. *The American Economic Review*, *84*(4), 753–771.

Naylor, R. L., Falcon, W. P., Rochberg, D., & Wada, N. (2001). Using El Nino/Southern oscillation climate data to predict rice production in Indonesia. *Climatic Change*, *50*(3), 255–265.

Nebraska Department of Agriculture. (2014). *Nebraska agriculture*. Retrieved from http://www.nda.nebraska.gov/publications/ne_ag_facts_brochure.pdf

Ray, D. K., Gerber, J. S., MacDonald, G. K., & West, P. C. (2015). Climate variation explains a third of global crop yield variability. *Nature Communications*, *6*. doi:10.1038/Ncomms6989

Reilly, J., Tubiello, F., McCarl, B., Abler, D., Darwin, R., Fuglie, K., ... Rosenzweig, C. (2003). US agriculture and climate change: New results. *Climatic Change*, *57*(1–2), 43–69.

Rosenzweig, C., & Parry, M. L. (1994). Potential impact of climate-change on world food-supply. *Nature*, *367*(6459), 133–138.

Scanlon, B. R., Faunt, C. C., Longuevergne, L., Reedy, R. C., Alley, W. M., McGuire, V. L., & McMahon, P. B. (2012). Groundwater depletion and sustainability of irrigation in the US High Plains and Central Valley. *Proceedings of the National Academy of Sciences of the United States of America*, *109*(24), 9320–9325.

Schlenker, W., & Roberts, M. J. (2009). Nonlinear temperature effects indicate severe damages to US crop yields under climate change. *Proceedings of the National Academy of Sciences of the United States of America*, *106*(37), 15594–15598.

Thompson, W., Gerlt, S., Campbell, J. E., Kueppers, L. M., Lu, Y. Q., & Snyder, M. A. (2017). A cost of tractability? Estimating climate change impacts using a single crop market understates impacts on market conditions and variability. *Applied Economic Perspectives and Policy*, *39*(2), 346–362.

Urban, D., Roberts, M. J., Schlenker, W., & Lobell, D. B. (2012). Projected temperature changes indicate significant increase in interannual variability of U.S. maize yields. *Climatic Change*, *112*(2), 525–533.

USDA NASS. *United States department of agriculture national agricultural statistics*. Retrieved from http://www.nass.usda.gov/

You, L. Z., Rosegrant, M. W., Wood, S., & Sun, D. S. (2009). Impact of growing season temperature on wheat productivity in China. *Agricultural and Forest Meteorology*, *149*(6–7), 1009–1014.

Can Sub-Saharan Africa feed itself? The role of irrigation development in the region's drylands for food security

Hua Xie, Nicostrato Perez, Weston Anderson, Claudia Ringler and Liangzhi You

ABSTRACT
This paper assesses the potential role of investments in irrigation in Sub-Saharan Africa in improving food security and self-sufficiency in the region. Focusing on the region's drylands, the study identifies a potential for expanded irrigated area of 6–14 million hectares (ha), depending on technology costs and other factors. Linkage of these results with a global agricultural trade model shows that accelerated irrigation investment can effectively reduce growing food import dependency from 54% under a business-as-usual scenario to a much smaller 17–40%; and can also reduce the population at risk of hunger and child under-nutrition.

Introduction

Sub-Saharan Africa has long been beset with food insecurity. Although progress has been made over the past few decades, the region is still faced with great challenges to produce sufficient food (Sasson, 2012; Rosen, Meade, Fuglie, & Rada, 2016). The latest *The State of Food Security and Nutrition* report (FAO, IFAD, UNICEF, WFP & WHO, 2017) finds that 23% of the population suffers from under-nutrition; and that food insecurity worsened, by almost 3 percentage points from 2014 to 2016.

High and growing dependency on imported food is another indicator for the worsening food security situation in the region. Crop yields of major staples in Sub-Saharan Africa lag substantially behind those of other regions of the world (FAOSTAT, 2016). This has resulted in high dependency on net food imports by many countries in the region. Among major countries in Sub-Saharan Africa, only Zambia achieved self-sufficiency of cereal supplies; in several other countries, such as Angola, Mauritania and Namibia, more than 50% of domestic needs for staple cereals are met through imports. Moreover, although there were increases in domestic production, improvements in cereal production were outstripped by population growth and increased demand by those benefiting from economic growth and moving to cities. As a result, the cereal

The opinions expressed and arguments employed in this publication are the sole responsibility of the authors and do not necessarily reflect those of the OECD or of the governments of its Member countries.

Supplemental data for this article can be accessed at https://doi.org/10.1080/02508060.2018.1516080

import dependency increased from 17% in 2000–02 to 19% in 2011–13 (FAOSTAT, 2016). The lack of self-sufficiency exposes the region to volatile global agricultural markets.

Ameliorating the food security situation in Sub-Saharan Africa requires the deployment of more effective technologies, institutions and policies in the agriculture and related sectors. These have been discussed in many studies (Chen et al., 2011; Devereux, 2016; Garrity et al., 2010; Rosegrant, Cline, Li, Sulser, & Valmonte-Santos, 2005; Rosegrant et al., 2017; Runge, Senauer, Pardey, & Rosegrant, 2003; Van Ittersum et al., 2016; Webber, Gaiser, & Ewert, 2014). A promising option is to expand irrigated agriculture. Food production is intrinsically linked to access to water for productive uses, and irrigated agriculture is overall more productive than rainfed agriculture. Globally, it is estimated that irrigated agriculture accounts for only 20% of cropland, but contributes 40% of total production (FAO, 2016). In Sub-Saharan Africa, crop production is predominantly rainfed; less than 5% of cropland is irrigated (FAO, 2016). Irrigation could well be an important means to help improve agricultural productivity and reduce food insecurity and import dependency in the region.

However, important knowledge gaps remain as to the specific gap that irrigation can and cannot fill in the complex food insecurity picture in the region. For example, irrigation investment is a costly and complex endeavour. The feasibility of irrigation expansion is influenced by many factors, both biophysical and socioeconomic. In two foresight analyses on food security in Sub-Saharan Africa noted above (Rosegrant et al., 2005; Van Ittersum et al., 2016), the expansion rates of irrigated agriculture were largely introduced as exogenous variables into the analyses. Their values were chosen to reflect trends of historical investment rather than irrigation development potential over the region in the future. This paper contributes to closing this gap by using an integrated biophysical and economic modelling approach to assess quantitatively the irrigation development potential in Sub-Saharan Africa and linking this investment with changes in food security and food import dependency.

The study focuses on the dryland zones of Sub-Saharan Africa where hunger and under-nutrition are most severe (Azzarri, Bacou, Cox, Guo, & Koo, 2016). Dryland areas here refer to arid, semi-arid and dry sub-humid zones classified using the aridity index (see Appendix I in the supplemental data online). They cover 43% of total land area (13.9 million km^2), 70% of cropland and 66% of cereal production in Sub-Saharan Africa and are home to about 425 million people, or half the region's population. Given the importance of dryland zones in the region's food production and the higher levels of food insecurity and under-nutrition in these areas, they have received special attention in recent analyses and provide particular promises for irrigation investment (Cervigni & Morris, 2016).

Data and methods

The methodological framework applied is presented schematically in Figure 1. The study consists of two analyses. The first is a strategic planning analysis, in which we estimate irrigation development potential in Sub-Saharan Africa. The strategic planning analysis starts with pre-suitability mapping, which uses multiple environmental suitability criteria to score the suitability of irrigation development potential of land pixels

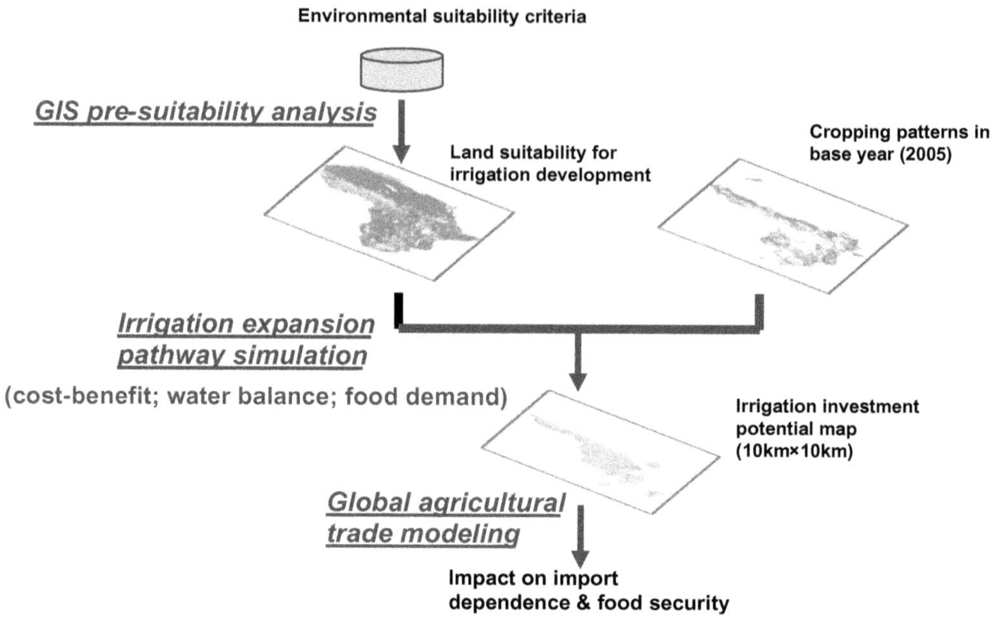

Figure 1. Methodological framework for irrigation development impact assessment.

in Sub-Saharan African countries on a 5 arc-minute grid. The pre-suitability mapping was followed by simulations, using a rule-based algorithm, guided by the calculated land suitability score, to simulate pathways of irrigation development under a series of constraints, including water availability, demand for irrigated crops and cost–benefit analysis of irrigated crop production. The results are identified irrigation development potential.

The second is a simulation analysis that incorporates the estimates of irrigation development potential in dryland zones in the International Model for Policy Analysis of Agricultural Commodities and Trade (IMPACT) (Robinson et al., 2015). IMPACT is a global, partial equilibrium agricultural sector model that has been widely applied to examine global and regional food security issues (Flachsbarth et al., 2015; Ringler, Zhu, Cai, Koo, & Wang, 2010; Rosegrant, Zhu, Msangi, & Sulser, 2008). The linkage between the two modelling systems allows one to examine broader-scale impacts of accelerated irrigation development on food availability, food security and net food trade.

Large-scale irrigation (LSI) and small-scale irrigation (SSI)

In this assessment, we distinguish between LSI and SSI. SSI refers to small irrigation schemes developed to harvest water resources to augment crop production under private ownership of smallholders, which could be fed by groundwater or surface water; LSI is associated with the construction of reservoirs with large storage capacities and is generally publicly financed. The analysis of potential for LSI development requires prior knowledge of locations and storage capacities of those reservoirs. To identify dams, we adapted a dam inventory across Africa provided by the World Bank. The inventory includes 680 dams, of which 373 have a capacity > 50 million m^3. Among the large dams, 253 are operational, while 120 dams are planned or slated for rehabilitation and were included in the study (see Appendix II in the supplemental data online). We assumed that no additional irrigated area can be linked to existing dams, as the area around dams in operation might already be fully exploited. The analysis also did not include barrages or run-of-the-river schemes. Both assumptions led to a lower-bound estimate of irrigation potential in the dryland zones of Sub-Saharan Africa.

Environmental suitability for pre-suitability mapping analysis

The environmental suitability criteria used in the *ex-ante* mapping for SSI and LSI are shown in Table 1. These criteria were determined by expert consultation. The sources of input data are shown in Table 2. The pixel size for the mapping of suitable areas is 0.5 × 0.5 km. A linear weighting scheme was used with all criteria weighted equally. The SSI suitability score of each pixel was calculated as:

$$SSI\ suitability\ score\ (0 \sim 100) = [S_1 + \max(S_2 + S_3) + S_4 + S_5]/4 \tag{1}$$

Table 1. Criteria for *ex-ante* suitability analysis.

Criteria	Range of parameter	Range of score
(a) *Small-scale irrigation (SSI)*		
Slope (S_1)	0–10%	S_1: 100–0
Distance to surface water (S_2)	0–5 km	S_3: 100–0
Groundwater depth (S_3)[a]	0–250 m	S_2: 100–0
Travel time to market (S_4)	0–3 h	S_4: 100–0
Distance to existing irrigation (S_5)	0–10 km	S_5: 100–0
(b) *Large-scale irrigation (LSI)*		
Slope (S_1)	0–10%	S_1: 100–0
Distance to main channel of river downstream of the dam (S_2)	0–5 km	S_2: 100–0
Distance to existing irrigation (S_3)	0–10 km	S_3: 100–0

Note: [a]No groundwater data were available for Madagascar, and so this scoring criterion was dropped for that country.

Table 2. Input data for *ex-ante* suitability analysis.

Criteria	Data set
Topography (slope)	Shuttle Radar Topography Mission (SRTM) elevation data[a]
Groundwater accessibility	British Geological Survey (BGS) quantitative groundwater map for Africa[b]
Distance to perennial surface water	Global Lakes and Wetlands Database[c]; V-Map Perennial Streamlines data set[d]
Proximity to existing irrigation	Food and Agriculture Organization (FAO) Sirte irrigation map "Water for Agriculture and Energy in Africa"[e]
Market access	Nelson's global travel time[f]
Urban extent	Urban Extents Grid, the Global Rural–Urban Mapping Project, version 1 (GRUMPv1)[g]
Protected area	World Database on Protected Areas (WDPA)[h]

Sources: [a]See https://cgiarcsi.community/data/srtm-90m-digital-elevation-database-v4-1/.
[b]Seehttp://www.bgs.ac.uk/research/groundwater/international/africanGroundwater/maps.html/.
[c]Seehttps://www.worldwildlife.org/pages/global-lakes-and-wetlands-database/.
[d]National Imagery and Mapping Agency (1997).
[e]See http://www.fao.org/nr/water.
[f]See http://forobs.jrc.ec.europa.eu/products/gam/.
[g]See http://sedac.ciesin.columbia.edu/data/collection/grump-v1/.
[h]Seehttps://www.iucn.org/theme/protected-areas/our-work/world-database-protected-areas/.

and the LSI suitability score is calculated as:

$$LSI\,suitability\,score(0\sim100) = [S_1 + S_2 + S_3]/3 \tag{2}$$

where S_1–S_5 are the scores for each suitability criterion (Table 1). Existing irrigated areas, urban areas and protected areas (Table 2) were removed following the pre-suitability analysis before assessing potential irrigation expansion. Pixels that do not meet slope and travel time ranges were also excluded. Moreover, for LSI, additional topographical analysis was conducted to delineate potential irrigation command areas associated with each dam. Irrigation from these dams was assumed to occur by gravity with command areas up to 200 km downstream of the dam (with the exception of Fomi Dam, which is being planned for multiple purposes, including irrigation up to 600 km downstream); and with water elevation heads of reservoirs of 10–20 m. The suitability criteria in Table 1(b) were applied to the delineated command areas.

For more details on the data and scoring schemes used in this study, see Appendix III in the supplemental data online. The results of the pre-suitability analysis were aggregated into 5 arc-minute areas (approximately 10 × 10 km) by averaging to inform the second step of the simulation analysis.

Simulation of irrigation expansion pathways

The simulation in this step of the analysis involved examining the likelihood of irrigation adoption in each pixel on the 5 arc-minute grid. The Spatial Allocation Model (SPAM) (You, Wood, Wood-Sichra, & Wu, 2014) was used as the base map for simulating irrigation expansion. SPAM 2005 provides spatially disaggregated estimates of cultivated area and yields for 42 crops in 2005 over the world under rainfed and irrigated conditions.

A complete description of the simulation algorithm is provided in Appendix IV in the supplemental data online. The algorithm was applied to each country in Sub-Saharan Africa to determine the irrigation development potential in that country. The design of the simulation algorithm used the following assumptions:

- The adoption of irrigation occurs in a sequence according to the suitability ranking of the pixels in a country. Irrigation first expands to the pixel with the highest suitability score, followed by the pixel with the second highest score etc.
- We first simulated the development of LSI in proximity of planned and to be rehabilitated dams, followed by an assessment of SSI in the remaining areas.
- Given the adverse impacts on forest cover and biodiversity of expanding agricultural land in Sub-Saharan Africa, pixels with existing rainfed croplands were first converted to irrigation, and only when irrigation remains sufficiently profitable were areas not yet cultivated converted.
- Internal rates of return (IRR) were used as a criterion for irrigation investment decisions, that is, irrigation investments will only be made if the IRR of irrigated crop production in the pixel is greater than a predefined level. The IRR calculation is based on the irrigated crop mix of the 5 arc-minute pixel.

The adoption of irrigation can be accompanied by a change in crop mix. We considered this important issue as follows. Irrigation development leads to double cropping with key second-season crops being maize, rice, wheat and vegetables. We allow for a larger set of candidate irrigated crops for the rainy season, including groundnuts, maize, millet, potatoes, sorghum, sugarcane, sweet potatoes, rice, vegetables and wheat. Second, farmers tend to plant high-value crops under irrigation given the higher input costs of irrigated agriculture. We reflect this by assuming that in each season the cultivated area of each crop is proportional to its net profitability (see equations (A-1) and (A-2) in Appendix IV in the supplemental data online), subject to meeting food demand needs in the country (see also below).

The potential for irrigation expansion is also constrained by the availability of renewable water resources and projected demand for irrigated crops. Water availability was evaluated for each reservoir (in the LSI analysis) and at the river basin level (for SSI analysis). Irrigation expansion within the potential command area of a reservoir or in a basin stops when the irrigation water supply capacity of the reservoir is reached or renewable water resources of the basin allocated to irrigation are fully used. The food demand constraint was applied to each individual irrigated crop at the national level. A crop is removed from the list of candidate irrigated crops or remaining simulations when the projected domestic demand for that crop by 2050 is fully met through increased irrigated production.

Irrigation costs and sensitivity analysis

Three levels of irrigation costs (low, medium and high) were considered in the sensitivity analysis, with ranges of US$8000–30,000/ha for LSI and US$3000–6000/ha for SSI (Table 3) based on an expert panel review of irrigation investment costs. The capital cost figures in Table 3 refer to the capital investment cost per investment cycle, and the operation and maintenance (O&M) cost figures are annual. While we do not differentiate between the costs of individual technologies, sensitivity analysis allows one to describe a range of technology cost options that can then be linked to various individual technologies.

We also implemented sensitivity analysis for two alternative IRR, 5% and 12%, which are rates typically considered by multilaterals in irrigation investment decisions, resulting in a total of six sensitivity analyses or alternative irrigation development pathways. To facilitate the discussion, we refer to the scenario with medium irrigation costs and IRR > 5% as the baseline scenario in the discussion, unless otherwise noted.

Table 3. Cost assumptions under three cost scenarios (US$/ha).

	Low		Medium		High	
	Capital[a]	O&M[b]	Capital	O&M	Capital	O&M
LSI	8,000	800	12,000	1,200	30,000	3,000
SSI	3,000	100	4,500	125	6,000	150

Note: LSI, large-scale irrigation; O&M, operation and maintenance; SSI, small-scale irrigation.
[a]costs per investment cycle; [b]costs per annum.

Assessment of impacts of irrigation expansion on food security and agricultural trade

IMPACT incorporates country-specific 'rates of irrigation development' in its 'business-as-usual' (BAU) scenario of agricultural development for 115 countries and regions, including Sub-Saharan Africa. These rates have been determined based on historic trends and expert assessment of likely increases in investment in the future. The strategic planning analysis in this study finds a substantially higher biophysical and socioeconomic potential for irrigation development, particularly through irrigation that is developed by individual farmers themselves than IMPACT BAU suggests. Table 4 compares irrigation development under the IMPACT BAU scenario with the rates of the dryland baseline scenario (medium irrigation costs plus IRR > 5%). We incorporated the faster irrigation expansion rates in the drylands of Sub-Saharan Africa into IMPACT and computed a series of food security indicators to show to what extent accelerated irrigation development in the dryland zones can support food security in the region. The calculated food security indicators include the number of undernourished children, the population at risk of hunger and net trade of major irrigated crops.

Results

Estimated irrigation development potential

The estimated irrigation development potential in Sub-Saharan Africa under the baseline scenario (medium irrigation costs and IRR > 5%) is 18 million ha across all land classes of Sub-Saharan Africa. A map displaying the identified sites with irrigation development potential under this scenario is included in Figure 2. Most of the potential, 14.8 million ha or 84% of the total potential, is for SSI. The potential for LSI, which is largely determined by the number of dams slated for rehabilitation or for construction over the next few decades, is much smaller and estimated at 3.2 million ha. Note that other estimates for irrigation development potential in Sub-Saharan Africa or Africa are available from a few previous studies. Results from different studies may vary substantially due to differences in approaches and data used. While the estimated irrigation development potential in this study is larger than the trends from historical investment reflected in IMPACT BAU, it is at the lower end compared with those estimates established using only a water balance approach (e.g., Altchenko & Villholth, 2015).

Table 4. Irrigated area change rates (%/year), IMPACT 'business as usual' (BAU) scenario and African drylands baseline (medium irrigation costs plus internal rates of return (IRR) > 5%) scenario.

	IMPACT BAU	Dryland baseline
Sub-Saharan Africa	1.66	3.02
Central Africa	2.61	5.30
Eastern Africa	0.10	1.89
Southern Africa	2.82	3.17
Western Africa	2.16	3.86

Note: IMPACT, International Model for Policy Analysis of Agricultural Commodities and Trade.

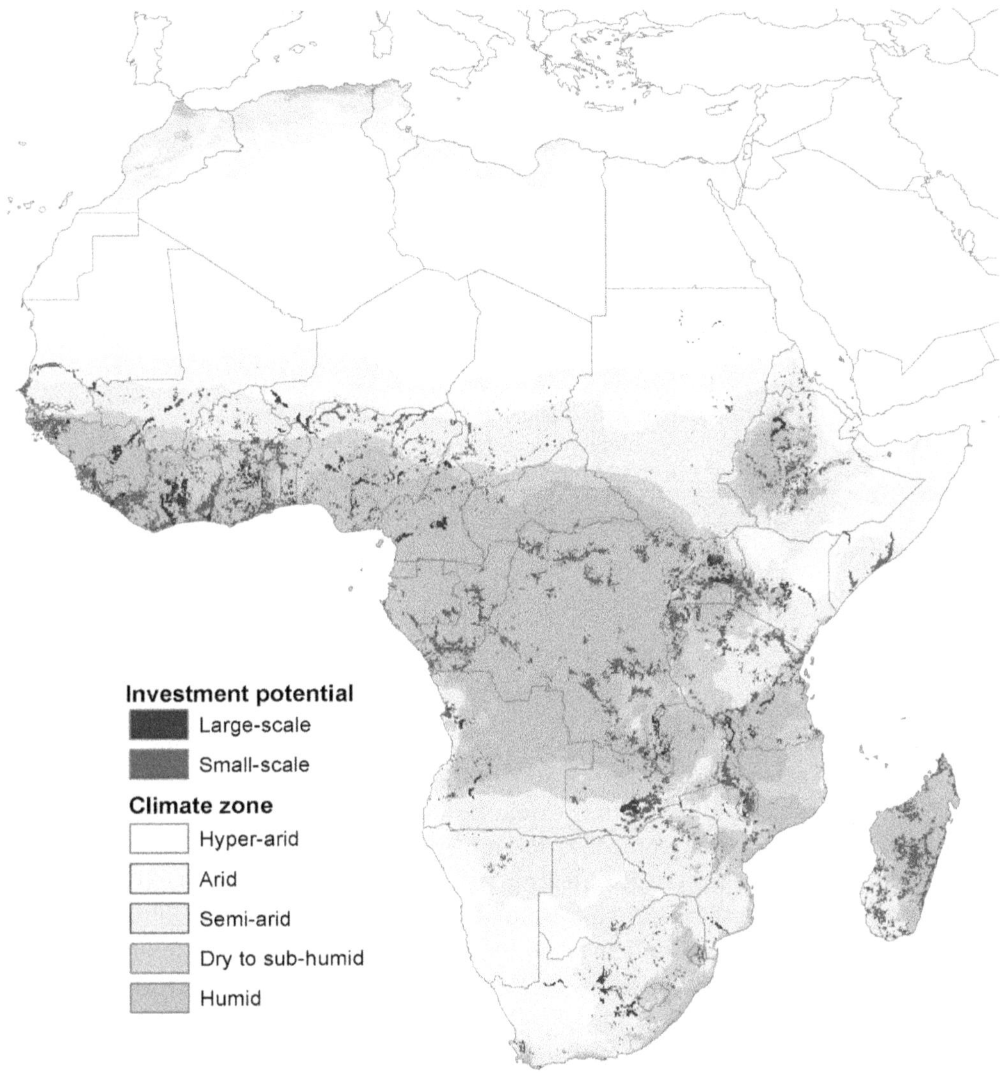

Figure 2. Estimated sites with irrigation development potential in Sub-Saharan Africa.

This is not surprising since we embraced a more integrated approach in our assessment considering cost–benefit analyses as well as the demand for irrigated crop production. The irrigation development potential reported here also serves as an update to the estimate reported by You et al. (2011), which was established in a similar conceptual methodology framework. More recently available input data and a new implementation approach were used in this study.

Table 5 summarizes the irrigation potential by land class. The combined LSI potential across the three dryland zones (arid, semi-arid and dry sub-humid) is 1.6 million ha, or about 50% of the total estimated LSI potential. The share of irrigation potential in dryland zones is larger for SSI: 9 million ha, or 60% of the total SSI potential in Sub-Saharan Africa.

Table 5. Estimated irrigation development potential in Sub-Saharan Africa by land class, medium irrigation cost and >5% internal rates of return (IRR) (ha, thousands).

Class	Large scale	Small scale
Hyper-arid	82	15
Arid	198	403
Semi-arid	686	5,303
Sub-humid	719	3,369
Humid	1,525	5,760
Total	3,210	14,850

Table 6. Distribution of estimated irrigation development potential in Sub-Saharan African drylands by sub-region, under medium irrigation cost and >5% internal rates of return (IRR) (ha, thousands).

Sub-region	Large scale	Small scale
East	326	2,358
Central	87	180
Southern	389	2,159
West	801	4,378
Total	1,603	9,075

Table 6 further summarizes the distribution of estimated area with irrigation potential in dryland zones by sub-region. A total of 10.7 million ha of new irrigation development is feasible, most of which (9 million ha) would be from SSI. Figure 3 ranks the irrigation development potential in dryland zones at country level. Among the four sub-regions in Sub-Saharan Africa, the largest potential for irrigation expansion in dryland zones is in West Africa, where approximately half the total potential is located. The LSI development potential in West Africa's dryland zone amounts to 0.8 million ha, and 4.4 million ha for SSI. The potential for irrigation expansion for East and Southern Africa is lower. The two regions contain 20% and 24% of the LSI development potential (0.3 and 0.4 million ha respectively), and 26% and 24% of the SSI potential (2.4 and 2.2 million ha respectively) in African drylands respectively. The estimated irrigation potential in drylands was lowest in Central Africa, at 0.1 million ha for LSI and 0.2 million ha for SSI, in part because this is the region with the lowest share of drylands in the region.

At the country level, Nigeria in West Africa is the country with the largest estimated potential for irrigation development (LSI plus SSI), accounting for almost half the total estimated irrigation potential in Sub-Saharan African drylands, at 2.5 million ha. Countries ranking second to fourth on the chart are Tanzania, Kenya and Malawi. They boast irrigation potential of more than 0.5 million ha. Other countries with significant irrigation development potential in drylands include Madagascar (0.49 million ha), Ghana (0.46 million ha), Senegal (0.39 million ha), Ethiopia (0.38 million ha), Zambia (0.37 million ha), Somalia (0.36 million ha) and Burkina Faso (0.33 million ha). An additional 12 countries have irrigation potential of between 0.1 and 0.3 million ha.

Both changes in capital cost and IRR affect values of final irrigated area potential. Figure 4(a,b) presents the estimated LSI and SSI potential in the drylands of Sub-Saharan Africa under all scenarios graphically. When the IRR cut-off value is held

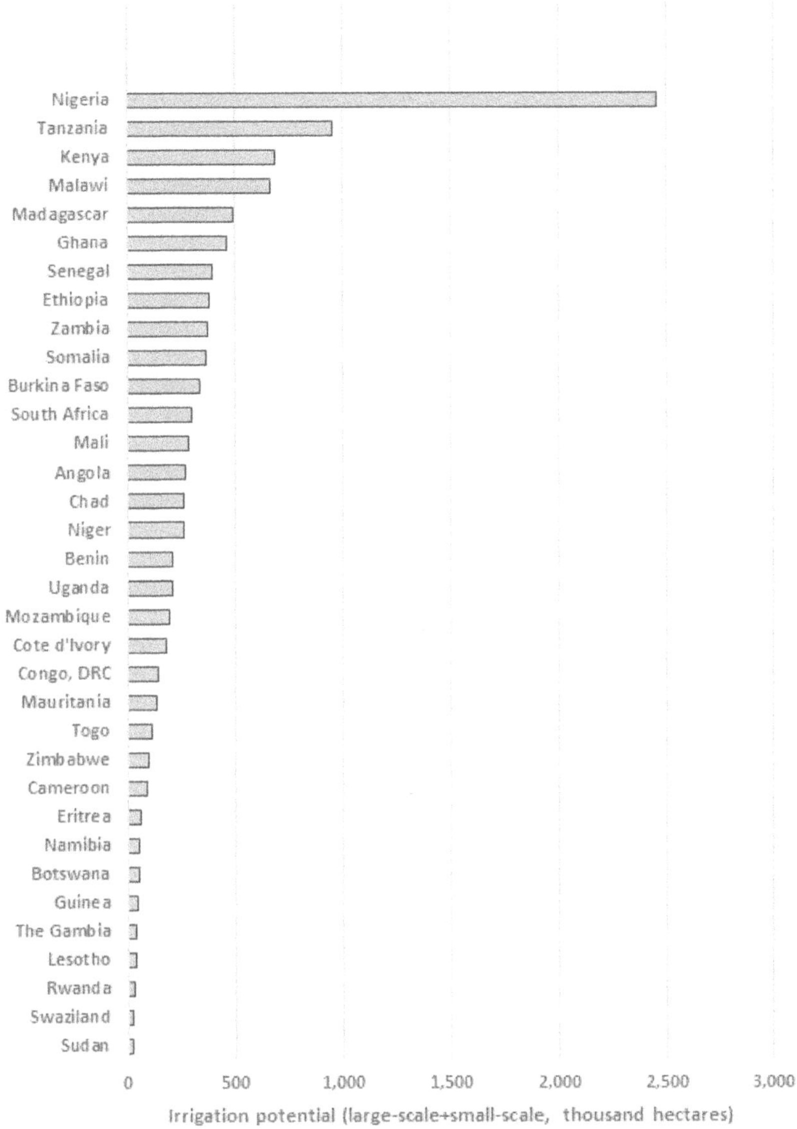

Figure 3. Estimated irrigation development potential in dryland zones by country under the baseline scenarios (medium irrigation costs and internal rates of return (IRR) > 5%) (no irrigation development potential on drylands was identified in Equatorial Guinea, Gabon, Liberia and Sierra Leone; countries with identified irrigation development potential in dryland zones but fewer than 20,000 ha are not shown: Burundi, 7000 ha; Djibouti, 4000 ha; Guinea-Bissau, 2000 ha; and Central African Republic, 2000 ha.

constant, an increase in irrigation costs leads to a reduction in estimated irrigation potential, and vice versa. For instance, with an IRR cut-off of 5%, the potential for LSI and SSI rise from 1.6 and 9.1 million ha to 2.5 and 11.6 million ha respectively, as irrigation costs drop from a medium to the low-cost level. By contrast, if irrigation costs increase, the potential area expansion in dryland areas declines to 1 million ha for LSI and 6.2 million ha for SSI. The potential for area expansion also declines if the IRR cut-

(a) Large-scale

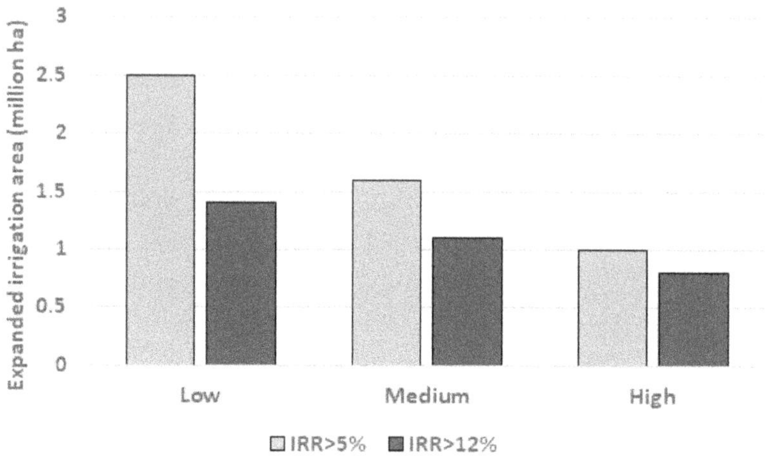

(b) Small-scale

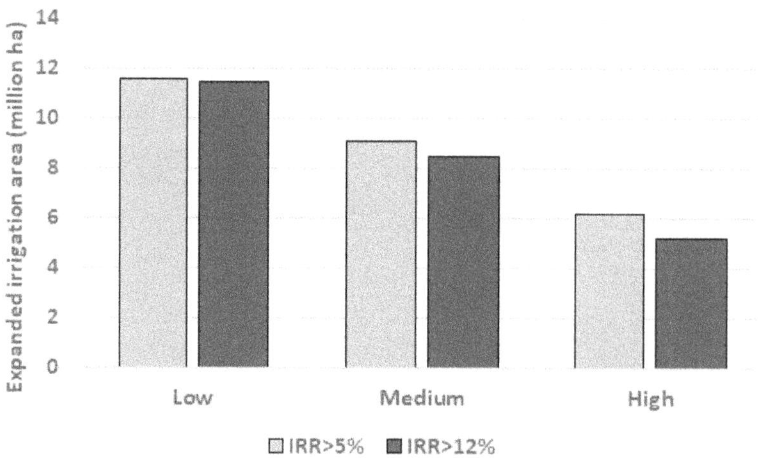

Figure 4. Potential irrigated area on Sub-Saharan Africa drylands under alternative internal rates of return (IRR) and cost assumptions (ha, millions), 2050 results.

off is raised. Under low irrigation cost assumptions, the irrigated area potential for LSI in African drylands declines from 2.5 to 1.4 million ha when the IRR cut-off increases from 5% to 12%, a decline of 43%. The decline is less severe under medium- and high-cost assumptions: under medium-cost assumptions, the area declines from 1.6 to 1.2 million ha when IRR increases from 5% to 12%, a 29% decline; and under high irrigation cost, the area declines from 1.01 to 0.85 million ha, a 16% decline. For SSI, on the other hand, which is generally more profitable, higher IRR are impacting area potential most under the high-cost scenario. At high cost, the potential for SSI in African drylands drops from 6.2 to 5.2 million ha, a 17% decline. Under the combined impacts of irrigation costs and IRR, the largest estimated irrigation development potential on drylands was obtained under the low-cost, low-IRR scenario, which consists of 2.5 million ha for LSI and 11.6 million ha for SSI; and the lowest estimate for irrigation potential was reported under the high-cost, high-IRR scenario, which includes 0.8 million ha for LSI and 5.2 million ha for SSI.

Food security indicators

Tables 7 and 8 and Appendix V in the supplemental data online present the impact of expanded irrigation investment under different IRR and irrigation cost levels on a series of food security indicators. These indicators were estimated using the IMPACT model at country level and aggregated to sub-continental regions in Sub-Saharan Africa.

The impacts of expanded irrigation development on the number of people at risk of hunger are presented in Table 7. The IMPACT BAU scenario suggests that, by 2050, 272 million people will be at risk of hunger in Sub-Saharan Africa. More than one-third of this population is based in Central Africa (92 million people), followed by Eastern Africa (67 million), West Africa (57 million) and Southern Africa (approximately 56 million). Accelerated irrigation development reduces these numbers by up to 14–15 million people under the low-cost scenarios (12% and 5% IRR respectively) and by 6–11 million people under the medium-cost scenarios.

Projected impacts of irrigation expansion are largest in Western Africa both in terms of percentage and number of people – a 9.5%, or 5.4 million, decline in the number of people at risk of hunger under the 5% low-cost scenario. For Southern Africa, the respective figures are 7.5% and 4.2 million; they are 4.9% and 3.3 million for Eastern Africa; and 2.1% and 1.9 million for Central Africa.

We also find that child under-nutrition in Sub-Saharan Africa can be decreased by 0.58–1.71% (0.2–0.7 million children) due to accelerated irrigation development in African drylands on top of baseline irrigation development – with the largest reduction in Southern Africa at 2.4%. Unsurprisingly, the scenario combining a 5% IRR and low irrigation costs (5% low) results in the largest impact, a 1.71% decline, followed by an IRR of 12% combined with low irrigation costs (12% low). The 12% IRR and high-cost scenario (12% high) results in the lowest decline of under-nourished children at 0.58% (for details, see Table A5-1 in the supplemental data online).

For all irrigation area-expansion scenarios in African drylands, net cereal (i.e., wheat, maize and rice) imports to the region decline, with decreases reaching as much as 68%, or 90 million tons (Table 8) from a baseline net import level of 133 million metric tons in 2050. The decline is largest for the 5% IRR low-cost scenario, followed by the 12% IRR low-cost scenario at 62% (82 million tons). At the sub-regional level, both the percentage and absolute decline in net imports is largest in Eastern Africa for the 5% IRR low-cost scenario, at 93% and 32 million tons. As such, accelerated investment in irrigation for both LSI and SSI in African drylands could reduce projected net cereal imports by more than two-thirds.

The cereal import dependency ratio is an important food security indicator. Combined with production data generated in IMPACT modelling, we calculate the cereal import dependency ratio under BAU and the scenarios with various accelerated irrigation investment rates. We find that accelerated irrigation investment could reduce the region's cereal import dependency ratio from 54% under the BAU

Table 7. Impact of expanded irrigation investment on food security–population at risk of hunger, 2050.

	Africa south of the Sahara		Eastern Africa		Central Africa		Southern Africa		Western Africa	
	Population at risk of hunger	Change from baseline	Population at risk of hunger	Change from baseline	Population at risk of hunger	Change from baseline	Population at risk of hunger	Change from baseline	Number of Undernourished Children	Change from baseline
Irrigation investment scenarios	Millions	%	Millions	%	Millions	%	Millions	%	Millions	%
Baseline	271.82	–	67.03	–	92.01	–	55.77	–	57.01	–
5% IRR and low cost	257.03	–5.44	63.73	–4.93	90.13	–2.05	51.60	–7.48	51.58	–9.53
5% IRR and medium cost	261.05	–3.96	65.26	–2.64	90.76	–1.36	52.85	–5.25	52.19	–8.45
5% IRR and high cost	267.19	–1.70	66.16	–1.29	91.18	–0.90	53.73	–3.66	56.12	–1.56
12% IRR and low cost	258.20	–5.01	64.20	–4.22	90.32	–1.83	51.89	–6.96	51.79	–9.15
12% IRR and medium cost	265.50	–2.33	65.67	–2.02	90.92	–1.18	53.18	–4.65	55.72	–2.26
12% IRR and high cost	268.25	–1.31	66.46	–0.85	91.34	–0.73	54.10	–3.01	56.35	–1.15

Note: IRR, internal rates of return.

Table 8. Impact of expanded irrigation investment on net trade, by commodity group, 2050(mt, thousands)[a]

	Africa south of the Sahara		Eastern Africa		Central Africa		Southern Africa		Western Africa	
Irrigation investment scenarios	Net trade	Change from baseline	Net trade	Change from baseline	Net trade	Change from baseline	Net trade	Change from baseline	Net trade	Change from baseline
(a) *Cereals (rice, maize, wheat)*										
Baseline	−132,614	–	−34,956	–	−9,019	–	−34,163	–	−54,475	–
5% IRR and low cost	−42,677	89,937	−2,583	32,373	−1,683	7,336	−6,309	27,854	−32,102	22,373
5% IRR and medium cost	−75,549	57,065	−13,132	21,825	−4,327	4,693	−23,103	11,060	−34,987	19,488
5% IRR and high cost	−92,207	40,407	−19,964	14,993	−5,439	3,580	−26,204	7,960	−40,600	13,874
12% IRR and low cost	−50,290	82,324	−5,837	29,119	−2,994	6,025	−8,332	25,832	−33,127	21,348
12% IRR and medium cost	−82,496	50,118	−15,855	19,101	−4,542	4,478	−25,205	8,958	−36,893	17,581
12% IRR and high cost	−99,710	32,904	−21,837	13,119	−5,657	3,363	−28,774	5,389	−43,442	11,033
(b) *Other grains (millet and sorghum)*										
Baseline	−1,714	–	3,945	–	1,605	–	−1,518	–	−5,747	–
5% IRR and low cost	1,182	2,896	3,454	−491	2,075	469	112	1,630	−4,459	1,288
5% IRR and medium cost	−723	991	1,781	−2,164	1,971	366	−444	1,074	−4,031	1,715
5% IRR and high cost	−2,128	−414	1,072	−2,873	1,813	207	−913	605	−4,100	1,647
12% IRR and low cost	498	2,212	2,782	−1,163	2,035	429	12	1,530	−4,330	1,416
12% IRR and medium cost	−1,384	330	1,375	−2,570	1,978	372	−807	710	−3,930	1,816
12% IRR and high cost	−2,516	−802	853	−3,092	1,810	205	−1,024	494	−4,156	1,591
(c) *Groundnuts*										
Baseline	−2,203	–	−990	–	−28	–	−195	–	−990	–
5% IRR and low cost	−574	1,629	−861	129	499	526	47	242	−258	732
5% IRR and medium cost	−1,082	1,121	−919	70	231	259	71	266	−465	526
5% IRR and high cost	−1,519	684	−956	34	56	83	40	235	−659	331
12% IRR and low cost	−810	1,393	−896	94	337	365	75	270	−327	664
12% IRR and medium cost	−1,185	1,018	−929	61	215	243	53	248	−524	466
12% IRR and high cost	−1,682	521	−991	−1	53	80	8	204	−752	238
(d) Vegetables										
Baseline	−35,702	–	−10,827	–	−9,129	–	−2,015	–	−13,731	–
5 % IRR and low cost	3,324	39,026	−6,696	4,131	−7,129	2,000	121	2,136	17,027	30,759
5% IRR and medium cost	3,339	39,040	−6,733	4,094	−7,166	1,963	181	2,197	17,055	30,787
5% IRR and high cost	1,612	37,314	−6,966	3,861	−7,157	1,972	94	2,109	15,641	29,373
12% IRR and low cost	2,990	38,692	−6,831	3,995	−7,137	1,992	135	2,150	16,823	30,555
12% IRR and medium cost	2,644	38,346	−6,938	3,888	−7,152	1,977	104	2,119	16,630	30,362
12% IRR and high cost	−142	35,560	−7,888	2,938	−7,240	1,889	−36	1,979	15,022	28,754

(*Continued*)

Table 8. (Continued).

Irrigation investment scenarios	Africa south of the Sahara		Eastern Africa		Central Africa		Southern Africa		Western Africa	
	Net trade	Change from baseline	Net trade	Change from baseline	Net trade	Change from baseline	Net trade	Change from baseline	Net trade	Change from baseline
(e) Root crops (potato and sweet potato)										
Baseline	−16,799	–	−13,925	–	−1,298	–	−9,444	–	7,867	–
5% IRR and low cost	35,226	52,025	874	14,799	−560	738	3,342	12,786	31,569	23,702
5% IRR and medium cost	29,037	45,836	528	14,453	−742	556	3,059	12,503	26,191	18,324
5% IRR and high cost	18,799	35,598	15	13,940	−392	905	2,056	11,499	17,121	9,254
12% IRR and low cost	33,188	49,988	526	14,451	−740	557	3,163	12,607	30,240	22,373
12% IRR and medium cost	25,857	42,657	179	14,103	−566	732	2,428	11,872	23,816	15,949
12% IRR and high cost	14,508	31,307	−176	13,748	−189	1,109	1,347	10,790	13,526	5,659
(f) *Sugar*										
Baseline	−8,908	–	−3,171	–	−695	–	2,401	–	−7,443	–
5% IRR and low cost	9,980	18,888	6,421	9,592	406	1,101	3,564	1,163	−411	7,032
5% IRR and medium cost	8,639	17,547	5,621	8,791	392	1,087	3,592	1,191	−965	6,478
5% IRR and high cost	2,524	11,432	1,665	4,836	317	1,012	3,806	1,405	−3,264	4,179
12% IRR and low cost	9,983	18,891	6,598	9,769	361	1,055	3,561	1,160	−537	6,906
12% IRR and medium cost	6,530	15,437	3,723	6,894	301	996	3,653	1,252	−1,147	6,296
12% IRR and high cost	682	9,590	1,091	4,262	340	1,035	3,651	1,250	−4,400	3,043

[a]Negative values for net trade relate to net imports, and positive values to net exports. A positive change means that 'net importers' are importing less, and 'net exporters' are exporting more. Negative changes represent 'net importers' importing more or 'net exporters' exporting less.

Note: IRR, internal rates of return.

scenario – which itself is a dramatic increase compared with today's dependency ratio – to 31% under the 5% IRR medium-cost scenario, 33% under the 12% IRR medium-cost scenario, 37% under the 5% IRR high-cost scenario, 40% under the 12% IRR high-cost scenario, 17% under the 5% IRR low-cost scenario, and to 20% under the 12% IRR low-cost scenario..

For other grains (millet and sorghum), which are generally not irrigated, the results are mixed – with reversals from net imports to net exports, increases in net imports and declines in net imports, depending on the irrigation scenario and sub-region. For the low-cost scenarios (5% low and 12% low), there are reversals from a small net import to a small net export situation. The highest gains of 2.9 and 2.2 million tons are projected under the two low-cost irrigation expansion scenarios.

For vegetables, the 2050 baseline suggests net imports of 36 million tons. With expanded irrigation development this situation reverses to net exports of as much as 3 million tons for the low IRR and low and medium irrigation-cost scenarios. Vegetables are generally higher value and thus less affected by higher costs or higher IRR. Even under the 12% high scenario, the trade position changes in absolute terms by 36 million tons, although the region remains a small net importer of 0.14 million tons of vegetables. Sub-regionally, Eastern and Central Africa remain net importers, but with large declines in imports – by as much as 38% and 23% respectively. Both Western and Southern Africa become net exporters – with Western Africa exporting up to 17 million tons of vegetables. This strong national production of vegetables under irrigation development likely also provides important nutritional benefits.

Finally, for sugar, Sub-Saharan Africa switches from a net import position in 2050 to a net export position under all accelerated irrigation expansion scenarios – with the highest net export levels – of 10 million tons – under the low-cost scenarios (both 5% and 12% IRR). This is achieved mostly through Eastern Africa's shift from a net-import position (of 3.2 million tons) to a net-export position (of 6.4 and 6.6 million tons), and through a reduction in net imports by Western Africa by as much as 7 million tons.

Discussion and conclusions

This paper assesses the potential role of irrigation investment in the dryland zone of Sub-Saharan Africa in improving food security and self-sufficiency of the region. In a strategic planning analysis, based on a series of assumptions, including environmental criteria for locating potential of LSI and SSI in African drylands, we find large areas for both sustainable and profitable SSI expansion, ranging from 5 to 12 million ha by 2050, with the final values depending on irrigation capital costs (low, medium or high) and acceptable levels of IRR (5% or 12% respectively). We also find substantial potential for increase in LSI of 0.8–2.5 million ha by 2050, all associated with a series of planned reservoirs and several dam rehabilitation projects.

Across dryland regions, the potential for irrigation expansion in this study is largest in West Africa, with approximately 5.2 million ha by 2050 (under the medium-cost, 5% IRR scenario, combining LSI plus SSI), accounting for 49% of all irrigation potential in dryland areas in Sub-Saharan Africa; this is followed by East Africa with 2.7 million ha, or 25% of the total potential, and Central Africa with 2.5 million ha, or 24% of the total potential. The potential for expansion of irrigated area is highly sensitive to capital costs of irrigation infrastructure. When costs of irrigation increase, net profitability of irrigation declines and the potential for irrigating staple crops is reduced, or not feasible/profitable. LSI and SSI area potential declines by more than half between low-cost, low-IRR and high-cost, high-IRR scenarios.

Through a further simulation analysis using the IMPACT model, we evaluated the consequences of accelerated irrigation investment by 2050 in the dryland areas of Sub-Saharan Africa in terms of food security. We conclude that there are important food security implications for the region from irrigation expansion. The most significant impact are changes in food import dependency. Accelerated irrigation expansion following the scheme directly affects the region's net trade position, resulting in a reduction of net cereal imports by as much as 90 million tons, a reduction of net vegetable imports by up to 39 million tons and a switch to a net export position for sugar. Accelerated irrigation investment reduces food import dependency for cereals from a staggering 54% under the baseline to 31% under the medium-cost and > 5% IRR scenario and a much lower 17% under the low-cost and > 5% scenario. Additionally, we found that the population at risk of hunger can and be reduced by 6–15 million people depending on the IRR and cost scenario and child under-nutrition in Sub-Saharan Africa can be decreased by 0.58–1.71% (0.2–0.7 million children).

In addition to those uncertainties and limitations that have been discussed, the study is also subject to additional limitations which merit mentioning below. First, we only reported assessment results under the IMPACT BAU scenario which does not consider climate change. Climate change influences values of various input variables of the planning analysis such as renewable water resources, irrigation water demand and therefore impacts the estimated irrigation development potential. Second, in the strategic planning analysis, we also used fixed crop prices and constant projections on future demands for irrigated crops, rather than deriving prices and demands through iteration with the global agricultural trade model. Finally, irrigation investment decisions are also influenced by local institutions and political favouritism. Extending the assessment to address these limitations is a topic of future research.

Acknowledgements

This paper was given at the workshop on Virtual Water in Agricultural Products: Quantification, Limitations and Trade Policy (Lincoln, Nebraska, USA, 14-16 September 2016), which was sponsored by the OECD Co-operative Research Programme: Biological Resource Management for Sustainable Agricultural Systems whose financial support made it possible for the author to participate in the workshop.

Disclosure statement

No potential conflict of interest was reported by the authors.

Funding

Part of this work was originally developed as a background paper for the Africa Drylands Study: Cervigni, R. and M. Morris, eds. (2016) Confronting drought in Africa's drylands: Opportunities for enhancing resilience. Africa Development Forum series. Washington, DC: World Bank. doi:10.1596/978-1-4648-0817-3, which received financial support from the Nordic Development Fund, the TerrAfrica Leveraging Fund, the United Nations Food and Agriculture Organization (FAO), the Program on Forests (PROFOR) Trust, and the CGIAR Research Program on Policies, Markets and Institutions (PIM). Participation in the conference leading up to this special issue paper was supported by the Organisation for Economic Co-operation and Development (OECD), Co-operative Research Programme: Biological Resource Management for Sustainable Agricultural Systems (CRP) while the writing of the paper was supported by the CGIAR Research Program on Water, Land and Ecosystems (WLE).

ORCID

Hua Xie http://orcid.org/0000-0001-8335-8160
Nicostrato Perez http://orcid.org/0000-0003-3030-4710
Weston Anderson http://orcid.org/0000-0003-3755-9943
Claudia Ringler http://orcid.org/0000-0002-8266-0488
Liangzhi You http://orcid.org/0000-0001-7930-8814

References

Altchenko, Y., & Villholth, K. G. (2015). Mapping irrigation potential from renewable groundwater in Africa–A quantitative hydrological approach. *Hydrology and Earth System Sciences Discussions*, *19*(2), 1055–1067.

Azzarri, C., Bacou, M., Cox, C., Guo, Z., & Koo, J. (2016). Subnational socio-economic dataset availability. *Nature Climate Change*, *6*, 115–116.

Cervigni, R., & Morris, M., editors, 2016. *Confronting drought in Africa's drylands: Opportunities for enhancing resilience*. Africa Development Forum series. Washington, DC: World Bank. doi:10.1596/978-1-4648-0817-3

Chen, X. P., Cui, Z. L., Vitousek, P. M., Cassman, K. G., Matson, P. A., Bai, J. S., ... Zhang, F. S. (2011). Integrated soil–Crop system management for food security. *Proceedings of the National Academy of Sciences*, *108*(16), 6399–6404.

Devereux, S. (2016). Social protection for enhanced food security in sub-Saharan Africa. *Food Policy*, *60*, 52–62.

FAO. 2016. AQUASTAT website. Food and Agriculture Organization of the United Nations (FAO). Website accessed on [2016/09/21].

FAO, IFAD, UNICEF, WFP and WHO. (2017). *The state of food security and nutrition in the world 2017. Building resilience for peace and food security*. Rome: FAO.

FAOSTAT. (2016). Suite of food security indicators. Retrieved September 21, 2016 from http://www.fao.org/faostat/en/#data/FS.

Flachsbarth, I., Willaarts, B., Xie, H., Pitois, G., Mueller, N. D., Ringler, C., & Garrido, A. (2015). The role of Latin America's land and water resources for global food security: environmental trade-offs of future food production pathways. *PloS One*, *10*(1), e0116733.

Garrity, D. P., Akinnifesi, F. K., Ajayi, O. C., Weldesemayat, S. G., Mowo, J. G., Kalinganire, A., ... Bayala, J. (2010). Evergreen Agriculture: A robust approach to sustainable food security in Africa. *Food Security*, *2*(3), 197–214.

Ringler, C., Zhu, T., Cai, X., Koo, J., & Wang, D. (2010). Climate change impacts on food security in sub-Saharan Africa. International Food Policy Research Institute (IFPRI) Discussion Paper, 1042.

Robinson, S., Mason-D'Croz, D., Sulser, T., Islam, S., Robertson, R., Zhu, T., ... Rosegrant, M. W. (2015). The international model for policy analysis of agricultural commodities and trade (IMPACT): Model description for version 3.

Rosegrant, M. W., Cline, S. A., Li, W., Sulser, T. B., & Valmonte-Santos, R. A. (2005). Looking ahead: Long-term prospects for Africa's agricultural development and food security. 2020 Vision Discussion Paper 41. Washington, DC: International Food Policy Research Institute.

Rosegrant, M. W., Sulser, T. B., Mason-D'Croz, D., Cenacchi, N., Nin-Pratt, A., Dunston, S., ... Willaarts, B., 2017. Quantitative foresight modeling to inform the CGIAR research portfolio. Project Report for USAID. Washington, DC: International Food Policy Research Institute (IFPRI). http://ebrary.ifpri.org/cdm/ref/collection/p15738coll2/id/131144

Rosegrant, M. W., Zhu, T., Msangi, S., & Sulser, T. (2008). Global scenarios for biofuels: Impacts and implications. *Review of Agricultural Economics*, *30*(3), 495–505.

Rosen, S., Meade, B., Fuglie, K., & Rada, N. (2016). International food security assessment, 2014–2024. *Economic Research*, *2014*, 2024.

Runge, C. F., Senauer, B., Pardey, P. G., & Rosegrant, M. W. (2003). *Ending Hunger in our Lifetime* (288 pp). Baltimore, Maryland: Johns Hopkins University Press.

Sasson, A. (2012). Food security for Africa: An urgent global challenge. *Agriculture & Food Security*, *1*, 1–16.

Van Ittersum, M. K., Van Bussel, L. G., Wolf, J., Grassini, P., Van Wart, J., Guilpart, N., ... Yang, H. (2016). Can sub-Saharan Africa feed itself? *Proceedings of the National Academy of Sciences*, *113*(52), 14964–14969.

Webber, H., Gaiser, T., & Ewert, F. (2014). What role can crop models play in supporting climate change adaptation decisions to enhance food security in Sub-Saharan Africa? *Agricultural Systems*, *127*, 161–177.

You, L., Ringler, C., Wood-Sichra, U., Robertson, R., Wood, S., Zhu, T., ... Sun, Y. (2011). What is the irrigation potential for Africa? A combined biophysical and socioeconomic approach. *Food Policy*, *36*(6), 770–782.

You, L., Wood, S., Wood-Sichra, U., & Wu, W. (2014). Generating global crop distribution maps: From census to grid. *Agricultural Systems*, *127*, 53–60.

Sustainability of aquifers supporting irrigated agriculture: a case study of the High Plains aquifer in Kansas

James J. Butler Jr., Donald O. Whittemore, B. Brownie Wilson and Geoffrey C. Bohling

ABSTRACT

The only means of moderating alarming depletion rates in many of the world's major aquifers is to reduce pumping. We apply a new water-balance approach to assess the impact of pumping reductions in a portion of the High Plains aquifer in the United States. Although projected aquifer responses to pumping reductions vary, practically achievable reductions would have a large impact throughout the area. These findings demonstrate that modest pumping reductions could greatly enhance prospects for groundwater-supported irrigation in the High Plains aquifer and similar aquifers elsewhere.

Introduction

Many of the world's major aquifers that support irrigated agriculture are being depleted at an alarming rate (Aeschbach-Hertig & Gleeson, 2012; Asoka, Gleeson, Wada, & Mishra, 2017; Scanlon et al., 2012). Food security, whether it is viewed from a local, regional, or global perspective, will be significantly affected if this depletion continues unabated (Dalin, Wada, Kastner, & Pum, 2017). In many semi-arid regions where surface-water supplies are limited, the most effective, if not only, means of moderating the rate of depletion and extending aquifer lifetimes is to reduce groundwater pumping. The key questions are 1) what levels of pumping reductions are required to bring these heavily stressed aquifers into a more sustainable condition, and 2) are such reductions practically achievable.

We will address these questions through an assessment of conditions in a portion of the High Plains aquifer (HPA) in the central United States, one of the world's largest and most important aquifers for irrigated agriculture. Products from groundwater-supported irrigation in the HPA are exported worldwide; significant reductions in these exports, and the virtual water transfers they represent, would have a significant impact on world food supplies (Marston, Konar, Cai, & Troy, 2015). Thus, it is of vital importance that we assess what can be done to bring the HPA and its groundwater-supported agricultural production into a more sustainable condition.

Supplementary data can be accessed here

The assessment will be performed using a recently developed water-balance approach, the Kansas Geological Survey – Water Balance Method (henceforth, KGS-WBM [Butler, Whittemore, Wilson, & Bohling, 2016]). The KGS-WBM, which requires data on annual water-level changes and annual groundwater pumping, can be used for assessments of areas ranging in size from hundreds to tens of thousands of square kilometers. It was developed for near-term (one to two decades) evaluations of seasonally pumped aquifers and appears particularly well-suited for use in semi-arid regions.

We will apply the KGS-WBM to the data-rich portion of the HPA that lies within the state of Kansas (Figure 1). Our focus will be on the pumping reductions required to significantly decrease the rate of water-level decline and the practical feasibility of those reductions. We will demonstrate the viability of the KGS-WBM using data from the first large-scale experiment in pumping reductions in the Kansas HPA. We will then apply the approach across the five groundwater management districts that overlie the Kansas HPA to develop a better understanding of what will be required to sustain this portion of the HPA and the globally important agricultural production it supports.

The KGS-WBM

The KGS-WBM is a lumped water-balance approach that applies to any area of an aquifer. In the simplest terms, an aquifer water balance can be expressed as:

$$Water\ Volume\ Change = Inflows - Outflows \tag{1}$$

Accurate quantification of the individual components making up aquifer inflows and outflows is, in principle, relatively straightforward, but rarely, if ever, achievable in practice. In response to this situation, the strategy of the KGS-WBM is to lump all aquifer inflows and all aquifer outflows except pumping into a single term, Net Inflow (Butler et al., 2016).

Equation (1) can then be restated as (Figure 2):

$$Water\ Volume\ Change = Net\ Inflow - Pumping \tag{2}$$

After rewriting in standard notation for an area of an unconfined aquifer like the HPA, Equation (2) becomes:

$$\Delta WL \times Area \times S_Y = I - Q \tag{3}$$

where ΔWL is the average water table change for the area [L]; S_Y is the specific yield, the parameter that characterizes how much water is released from or taken into aquifer storage as the position of the water table changes [-]; I is net inflow [L^3]; and Q is pumping [L^3]; with all quantities being defined on an annual time frame.

Many aquifers that support irrigated agriculture are pumped seasonally, i.e., groundwater is pumped for irrigation during the growing season with relatively little pumping during the rest of the year. Under those conditions and given that water-level measurements are taken long enough after the cessation of pumping that they display little sensitivity to variations in the timing of the irrigation season, Equation (3) can be rearranged and simplified to the following form (Butler et al., 2016):

$$\Delta WL = \frac{I}{Area \times S_Y} - \frac{Q}{Area \times S_Y} \approx b - aQ \tag{4}$$

where a and b are constants (= 1/(Area × S_Y) and I/(Area × S_Y), respectively). The assumptions required to move to the right-hand side of Equation (4) are that S_Y and I change little with time; as will be shown here, these assumptions appear appropriate for much, if not all, of the Kansas HPA.

Application of the KGS-WBM involves applying Equation (4) to plots of average annual water-table change versus annual pumping. Given the constants of the best-fit line, the water-table change resulting from a reduced level of pumping can be calculated using Equation (4). The pumping required to reach stable water levels (Q_{stable}) can be found by setting ΔWL to zero and rearranging Equation (4) (Butler et al., 2016):

$$Q_{stable} = I = \frac{b}{a} \tag{5}$$

The specific yield of an area can be found by rearranging the slope (a) constant in Equation (4):

$$S_Y = \frac{1}{(Area \times a)} \tag{6}$$

Application to the High Plains aquifer

The HPA is one of the world's largest and most productive aquifers and has supported critically needed agricultural production for decades. The aquifer extends over eight states in the High Plains region of the central United States and provides irrigation water primarily for alfalfa, corn, cotton, sorghum, soybean, and wheat (Smidt et al., 2016). The HPA is heavily stressed in its central and southern areas, where groundwater withdrawals have produced large water-level declines that threaten the viability of irrigated agriculture in those areas (McGuire, 2017; Scanlon et al., 2012). In the state of Kansas, the HPA is the most important source of water for western and south-central Kansas, supplying over 90% of the irrigation water used in the state (Buchanan, Wilson, Buddemeier, & Butler, 2015). This heavy reliance on groundwater has come at a steep price in western Kansas (Figure 1), as continuation of current decline rates will lead to resource depletion within a matter of a few decades or less over much of the area (Butler, Stotler, Whittemore, and Reboulet, 2013; Kansas Geological Survey, 2018). The impact of resource depletion on the agricultural productivity and economic vitality of the area would be great.

The management of water resources in Kansas is the responsibility of the Division of Water Resources in the Kansas Department of Agriculture (KDA). The KDA is assisted in the management of groundwater in the Kansas HPA by five groundwater management districts (GMDs) that overlie the vast majority of the aquifer in Kansas (Figure 1). The GMDs, which were formed in the 1970s to allow local input into management of groundwater resources, can propose management initiatives that are tailored to their circumstances as long as those initiatives fit within the framework of Kansas water law (Peck, 1980).

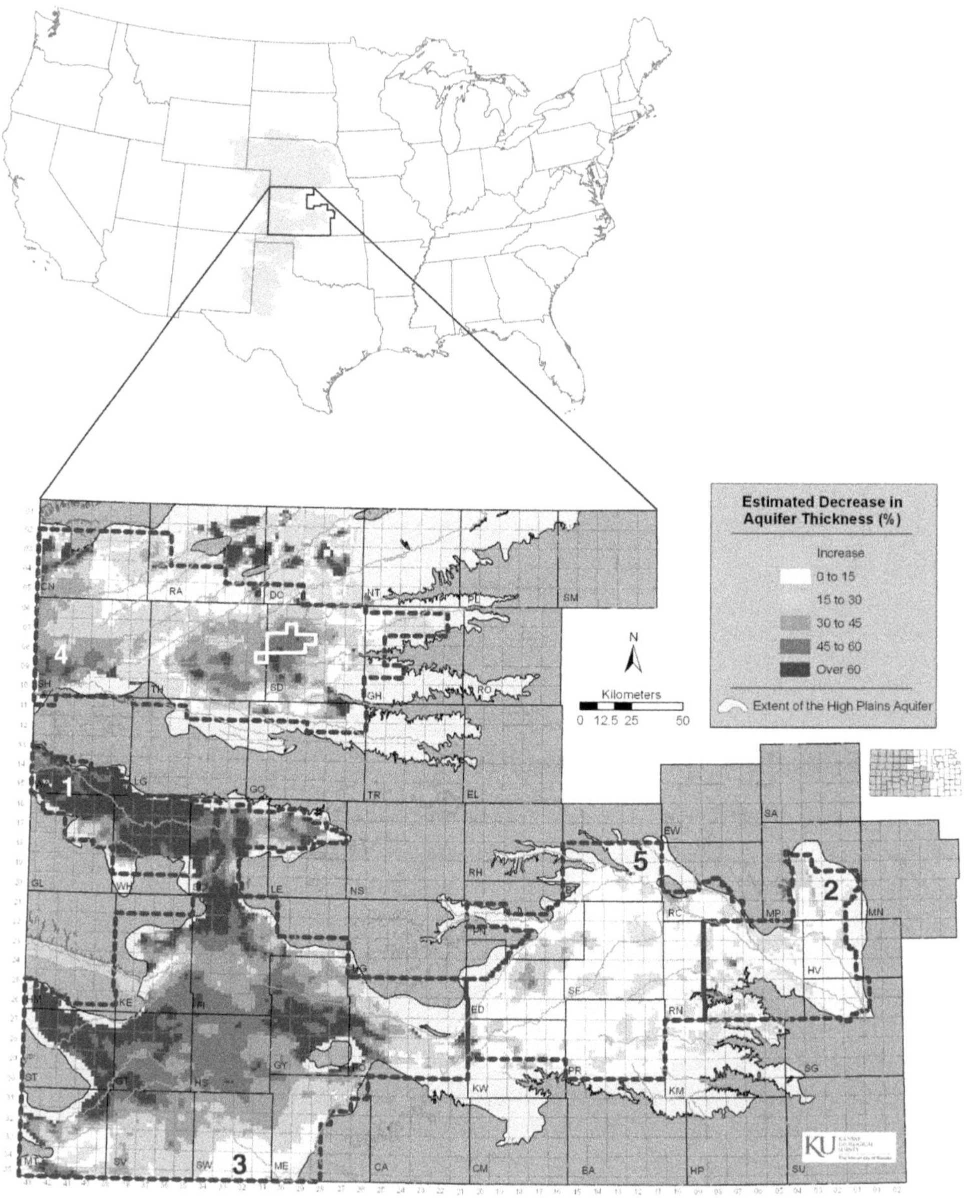

Figure 1. Location map of the High Plains aquifer (HPA) in the United States with an expanded view of the percent change in aquifer thickness from predevelopment to present for the HPA in Kansas. Predevelopment is defined as period prior to onset of widespread pumping for irrigated agriculture, which occurred between 1940 and the mid-1950s in most of the Kansas HPA; present is defined as average of 2015–2017 winter conditions. The areas of increase in the western third of the state are areas of thin saturated thickness that are of little practical importance. The areas comprising the groundwater management districts are enclosed by the dashed lines and numbered (black or white for clarity); the area comprising the Sheridan-6 LEMA is enclosed by the solid line (Kansas Geological Survey, 2018).

Given the scarcity of surface water in western Kansas, reductions in groundwater pumping are essentially the only means of moderating rates of water-table decline. Groundwater reductions, however, must be implemented over a large area to make a significant impact on regional decline rates. Such large-scale reductions require a management framework that makes them possible. In 2012, a new groundwater management option, the Local Enhanced Management Area (LEMA), was established in Kansas. The LEMA framework allows the development of locally generated management plans that are then supported by regulatory oversight (Kansas Department of Agriculture, 2013; Kansas Statutes Annotated 82a-1041, 2012). Recently, additional management options have been developed to further enhance the prospects for large-scale reductions in pumping (Groundwater Management District Act, 2017; Kansas Senate Bill 156, 2015).

The first LEMA was established in 2013 in a 256 km^2 area of Groundwater Management District #4 (GMD4) in northwestern Kansas (Sheridan-6 [SD-6] LEMA – Figure 1); this area had previously been identified as a highly stressed portion of the Kansas HPA. The LEMA goal was to reduce annual water use (all pumping of groundwater) by 20% from the average annual use for 2002–2012 (Northwest Kansas Groundwater Management District No. 4, 2016). The plan was initially adopted for five years. Participants were each allocated a total water volume for the five-year period and were free to choose how they distributed the water use over the period; the water volume calculation was the same for all irrigators (irrigated area multiplied by five times the allowable annual water use/unit area [approximately 28 cm]). Four years have been completed and participants recently voted to extend the plan for another five years (Kansas Department of Agriculture, 2017); an unused water volume equivalent to one year of average use can be transferred to the next five-year cycle.

The KGS-WBM is first applied to the SD-6 data to assess the capability of the method to reproduce the water-table response to pumping reductions. Following that validation demonstration, the approach is then applied to the Kansas HPA at

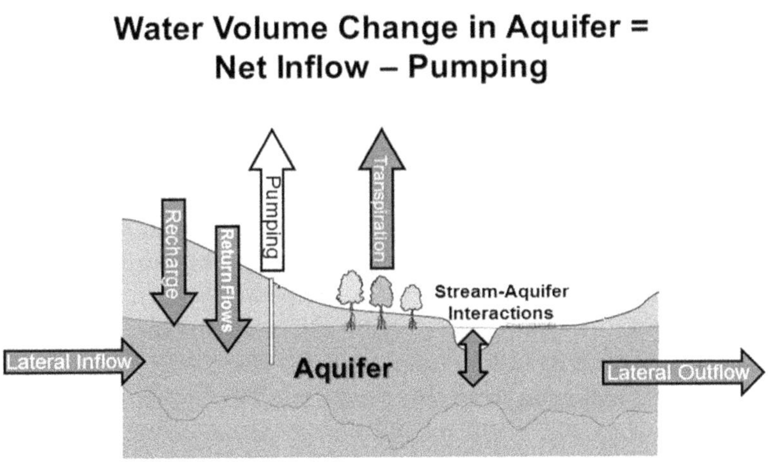

Figure 2. Schematic of the KGS – Water Balance Method. The dark arrows are examples of the components of the aquifer water balance that are included in the net inflow term; all components of the aquifer water balance save pumping are lumped into this term.

the scale of the GMDs (3,500–21,600 km^2) as a follow-up to an initial analysis presented in Butler et al. (2016). In all cases, the water-level data used in the analyses are from annual water-level measurements taken each winter, primarily in late December to early January, three to four months after cessation of irrigation pumping (Butler et al., 2016; Miller, Buchanan, & Brosius, 1999). The ΔWL value for a given well for a given year is the difference between the depth to water measurement that year and the following. The ΔWL value for an area for a given year is the average of those differences for that year. The pumping data are from the annually reported volumes for all non-domestic pumping wells in an area; as of 2013, over 95% of the non-domestic pumping wells in the Kansas HPA had totalizing flowmeters (Lanning-Rush, 2016). The annual pumping volume recorded by a flowmeter must be reported to the KDA annually and is subject to regulatory verification. The water-use data for a given year must be reported by March of the following year; those data are released, after quality-control checking, by KDA in late summer or fall of that year.

SD-6 validation demonstration

In the eleven years prior to the establishment of the SD-6 LEMA, groundwater levels were measured annually in late December or early January in six monitoring wells distributed across the area (Figure S1). Following the establishment of the LEMA, additional wells were added to the network in 2012–2015 (Figure S1). Prior to and following the establishment of the LEMA, water use has been reported annually for the approximately 195 irrigation wells (number varies slightly from year to year) in the area. In the following paragraphs, the application of the KGS-WBM to the SD-6 LEMA data is discussed; results are presented as the mean and 95% confidence interval (henceforth, CI) about it.

The plot of 2002–2012 (pre-LEMA period) ΔWL and Q data is approximately linear (solid line in Figure 3); application of the KGS-WBM yields an I of 21.5 × 10^6 m^3 (CI = 9.7–26.5 × 10^6 m^3). The average annual water use (Q_{av}) for 2002–2012 was 36.9 × 10^6 m^3. Thus, the pumping reduction required to achieve stable water levels is 42% (1-(I/Q_{av})), while the goal of the SD-6 LEMA was to reduce pumping by 20% (Q_{goal} = 29.5 × 10^6 m^3).

The plot of 2002–2016 (pre-LEMA and LEMA periods) ΔWL and Q data differs only slightly from that of the pre-LEMA period (dashed line in Figure 3), demonstrating the consistency of the pre-LEMA and LEMA data with the theory underlying the water-balance method. The Q_{av} for the LEMA period (2013–2016) was 23.7 × 10^6 m^3, a 36% reduction from Q_{av} for the pre-LEMA period, while the ΔWL_{av} for the LEMA period was 0.138 m/yr, a 76% reduction from that for the pre-LEMA period. Application of the KGS-WBM yields an I of 20.4 × 10^6 m^3 (CI = 15.9–23.3 × 10^6 m^3), revealing that the large reductions in Q_{av} and ΔWL_{av} had little impact on the net inflow estimate. Four growing seasons during the pre-LEMA period (2002–2004, 2012 – Table S1) were considerably drier than those during the LEMA period. Removing those four years, during which Q was much greater than the average for the period, and recalculating Q_{av} and ΔWL_{av} for the pre-LEMA years results in 33.6 × 10^6 m^3 and 0.419 m/yr, respectively. Thus, 29% and 67% reductions in the average annual pumping and water-level

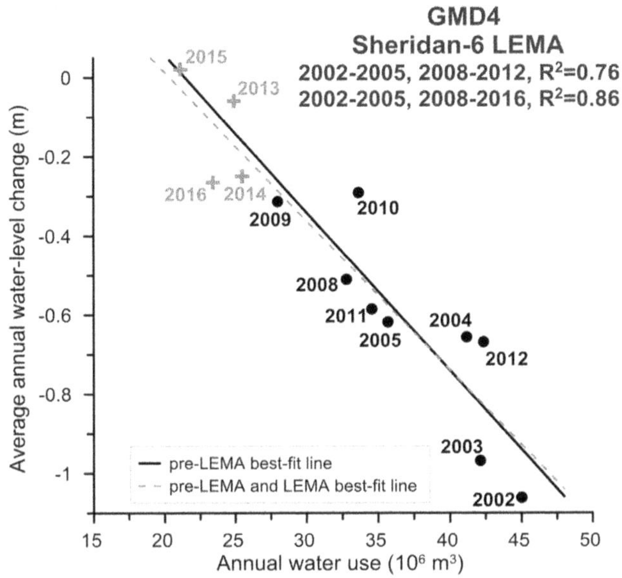

Figure 3. Average annual water-level change (ΔWL) versus annual water use (Q) plot for the SD-6 LEMA in GMD4. Solid line is best-fit line to 2002–2012 data ($\Delta WL = 0.856–0.0399Q$, $p < 0.003$) and dashed line is best-fit line to 2002–2016 data ($\Delta WL = 0.767–0.0376Q$, $p < 0.0001$). ΔWL for the circles is the average water-level change for the six wells measured every year from 2002 to 2012 (pre-LEMA wells); ΔWL for the pluses is the average for the pre-LEMA wells with two, three, and four additional wells (2013, 2014–15, and 2016, respectively; see Figure S1). Heavy snows delayed 2007 water-level measurements from early January to late February through early April, so ΔWL values for 2006 and 2007 are not used here (see Figure S1). Annual water use is sum of reported use from a maximum of 195 pumping wells (total varies slightly from year to year). The estimated uncertainty (one standard deviation) in ΔWL is ±0.19 m and that in Q is ±0.43% of plotted value (determined using methods described in Butler et al., 2016). Data are provided in Table S1.

decline rate, respectively, were achieved in the SD-6 LEMA for more similar climatic conditions.

This analysis demonstrates that 1) practically feasible reductions in pumping can have a large impact on water-level decline rates, and 2) the KGS-WBM can predict the response to pumping reductions in seasonally pumped aquifers using only water-level and water-use data from the pre-reduction period. This tool will now be applied to assess the impact of pumping reductions across the HPA in Kansas.

Application to the Kansas GMDs

In the following subsections, the KGS-WBM is applied to each of the Kansas GMDs. The ordering is from north to south and west to east, so that discussions of GMDs with similar geographic and hydrologic conditions are in close proximity (Figure 1). The emphasis is on the % reduction of average annual pumping required to attain stable water levels ($\Delta WL = 0$). In this case, stable water levels are defined in an areal average sense, i.e., the water-level change at any particular point may be negative or positive, but the spatial average of the changes across the district is zero. The analyses are performed for the 1996–2016 time period and, if shorter, the period for which over two thirds of the wells in the GMD were equipped with totalizing flowmeters. In the plots (supplementary materials) accompanying the following subsections, only the data points for the last five years and other years of interest are labelled to enhance plot clarity.

GMD4

GMD4 is a 12,623 km^2 area in semi-arid northwestern Kansas. A plot of ΔWL versus Q is approximately linear (solid line in Figure S2); Q_{stable}, which is calculated from the parameters of the best-fit line (Equation (5)), is 0.403 × 10^9 m^3. Thus, the average annual pumping from 1996–2016 (Q_{av} = 0.524 × 10^9 m^3) would need to be reduced by 23% (1-(Q_{stable}/Q_{av})) to achieve areally averaged stable water levels. The Q_{stable}/area for GMD4 is 0.032 m and the S_Y estimate is 0.053. The density of water use (Q_{av}/area) is 0.042 m, the lowest for the Kansas GMDs. Repeating the analysis for the period for which over two thirds of the wells were equipped with totalizing flowmeters (2008–2016 – dashed line in Figure S2) lowers Q_{stable} to 0.372 × 10^9 m^3 and increases the percent reduction to attain areally averaged stable water levels to 27%. In that case, the Q_{stable}/area for GMD4 is 0.029 m and the S_Y estimate is 0.063.

GMD1

GMD1 is a 4,734 km^2 area in semi-arid west-central Kansas. This district was the last of the GMDs to require installation of totalizing flowmeters. In 2010, 67% of the non-domestic pumping wells had totalizing flowmeters, as opposed to 99% in GMDs 3–5 (every non-domestic well with a water right in GMD1 was required to have a totalizing flowmeter as of 1 January 2014) (Kansas Department of Agriculture, 2011). Prior to installation of flowmeters, the district mostly relied on duration-of-pumping meters (duration meters) for annual estimates of water use. Because of temporal variability in flow during the irrigation season as well as lack of pumping-rate information, estimates from duration meters can have sizable error introduced by the assignment of a pumping rate to convert duration of operation to volume of water use. A plot of ΔWL versus Q for 1996–2016 found very low correlation, likely as a result of error in the annual pumping estimates (Figure S3). Isolating the analysis to the period for which at least two thirds of the wells had totalizing flowmeters (2010–2016) produced a near-linear plot. The Q_{stable} calculated from the parameters of the best-fit line to the 2010–2016 plot is 0.159 × 10^9 m^3. Thus, the average annual pumping from 2010–2016 (Q_{av} = 0.233 × 10^9 m^3) would need to be reduced by 32% to achieve stable water levels. The Q_{stable}/area for GMD1 is 0.034 m and the S_Y estimate is 0.085. The density of water use (Q_{av}/area) is 0.049 m.

GMD3

GMD3 is a 21,604 km^2 area in semi-arid southwest Kansas. Although the district had a large percent of totalizing flowmeters installed by the late 1990s (95% by 2000), it took some time before those flowmeters were reporting reliable data (L. Letourneau, Water Appropriations Program Manager, KDA, personal communication, 16 August 2016 and 15 November 2016). This is illustrated by a plot of ΔWL versus Q for 1996–2016 (Figure S4), the correlation is low and the reported annual water use for the first decade appears to be much greater than that for the second decade for similar climatic conditions. For example, climatic conditions (as characterized by the Standardized Precipitation Index (SPI) (McKee, Doesken, & Kleist, 1993; National Climatic Data Center, 2017) for 2015 were drier than those for 1996 and somewhat drier than those for 2004, and the absolute water-level change in 2015 was much greater than that in 1996 and somewhat greater than that in 2004. However, in marked contrast to the climatic and water-level change data, the reported pumping volumes for 1996 and 2004 were 28% and 14% greater, respectively, than that for 2015. Similarly, climatic conditions for 2000 were somewhat wetter than those for 2014 and the water-level changes were near equal, but the reported pumping for 2000 was 14% greater than that for 2014. Given the uncertainty in the reliability of the pre-2005 reported water use, the analysis was limited to the 2005–2016 period. The plot of ΔWL versus Q for 2005–2016 is near linear. The Q_{stable} calculated from the parameters of the best-fit line to the 2005–2016 plot is 1.570×10^9 m^3. Thus, the average annual pumping from 2005–2016 ($Q_{av} = 2.329 \times 10^9$ m^3) would need to be reduced by 33% to achieve stable water levels. The Q_{stable}/area for GMD3 is 0.073 m and the S_Y estimate is 0.056. The density of water use (Q_{av}/area) is 0.108 m, the highest in the Kansas GMDs.

GMD5

GMD5 is a 10,120 km^2 area in subhumid south-central Kansas. Despite the outlier (2007), a plot of ΔWL versus Q is approximately linear (solid line in Figure S5); the Q_{stable} calculated from the parameters of the best-fit line is 0.603×10^9 m^3. Thus, the average annual pumping from 1996–2016 ($Q_{av} = 0.621 \times 10^9$ m^3) would need to be reduced by 2.9% to achieve stable water levels. The Q_{stable}/area for GMD5 is 0.060 m and the S_Y estimate is 0.031. However, I for 2007 was considerably greater than that for the other years (41% larger according to the 1996–2013 analysis in Butler et al., 2016). Thus, the assumption of a constant I through time appears problematic for a dataset that includes 2007. The analysis was therefore repeated for the 1996–2006, 2008–2016 dataset (dashed line in Figure S5). In this case, the calculated Q_{stable} is 0.589×10^9 m^3 and a reduction of 6.1% from the average annual pumping for 1996–2016 is required to obtain stable water levels (Q_{stable}/area = 0.059 m, $S_Y = 0.035$). This reduction is over 110% larger than that calculated with the full dataset. Thus, recharge during infrequent high inflow years plays an important role in keeping the HPA in GMD5 close to a stable water-level condition. The density of water use (Q_{av}/area) is 0.061 m.

GMD2

GMD2 is a 3,543 km^2 area in subhumid south-central Kansas. A plot of ΔWL versus Q is approximately linear (Figure S6); the Q_{stable} calculated from the parameters of the best-fit line is 0.225×10^9 m^3. Thus, the average annual pumping from 1996–2016 ($Q_{av} = 0.223 \times 10^9$ m^3) could actually be increased by 0.9% and still result in stable water levels. The Q_{stable}/area for GMD2 is 0.064 m and the S_Y estimate is 0.030. The GMD2 relationship is much noisier than those for the districts to its west. In GMD2, the water table is shallower, the precipitation is greater, and stream-aquifer interactions play a more prominent role in the aquifer water balance. The assumption of a constant I may therefore not be as strong as in the districts to the west. As with GMD5, infrequent high inflow years (e.g., 2007 and 2013) appear to play an important role in keeping the

HPA in GMD2 close to a stable water-level condition. The density of water use (Q_{av}/area) is 0.063 m. Repeating the analysis for the period for which over two thirds of the wells were equipped with totalizing flowmeters (2003–2016) had a negligible effect on Q_{stable}.

Discussion and conclusions

The depletion of aquifers supporting irrigated agriculture and the ramifications of that depletion for food supplies is a pressing concern worldwide (Dalin et al., 2017; Hanjra & Qureshi, 2010; Marston et al., 2015). The large imbalance between aquifer inflows and outflows must be reduced to extend aquifer lifespans and maintain groundwater-supported agricultural production. Pumping reductions are often the only viable option for reducing this imbalance. A recently developed lumped water-balance approach, the KGS-WBM, has great potential for assessing the required levels of pumping reductions (Butler et al., 2016). Using the results from the first four years of a controlled demonstration of pumping reductions, we have shown that this approach can predict the response to pumping reductions in seasonally pumped aquifers using only water-level and water-use data from the pre-reduction period. Thus, we now have a theoretically sound, yet readily applicable, approach that can provide the water-resources community with reliable estimates of the impact of proposed pumping reductions.

The KGS-WBM was used here for a case study of the pumping reductions required to attain more sustainable conditions in the High Plains aquifer in Kansas. An analysis of conditions across the aquifer in Kansas indicates practically feasible pumping reductions would have a large impact on water-level decline rates. In the eastern portions of the region, there appears to be considerable potential for attaining stable water levels with small reductions in pumping. In south-central Kansas, reduction of the average annual pumping for 1996–2016 by 6% or less would lead to stable water levels for at least the near-term. In the portions of the HPA in semi-arid western Kansas, considerably greater reductions, 23–33%, would be required. Even in western Kansas, however, relatively modest reductions in pumping (10–20%) would lead to significant decreases in water-level decline rates. For example, in GMD3, Equation (4) can be used to show that a 15% reduction in the average annual pumping for 2005–2016 would lead to a 46% reduction in the average annual water-level decline (from the measured −0.622 m decline to the computed −0.338 m). Such a reduction in the average annual decline rate would significantly extend the lifespan of the aquifer.

This analysis produced three basic groupings of the GMDs in the Kansas HPA:

(1) GMDs 1 and 4 – these GMDs in semi-arid west-central and northwest Kansas have similar Q_{stable}/area values (0.029–0.034 m) with a somewhat greater spread in S_Y (0.053–0.085). The similarity in Q_{stable}/area is not unexpected given the similar relationships observed between water-level changes and climatic indices in these two districts (Whittemore, Butler, & Wilson, 2016). The density of water use is somewhat greater in GMD1 than in GMD4 (0.049 m vs 0.042 m) and the reduction required to attain stable water levels varies from 23 to 32%. However, limiting the GMD4 analysis to the period for which over two thirds of the wells

had totalizing flowmeters produced a reduction closer to that for GMD1 (27% versus 32%);

(2) GMD3 – this GMD in semi-arid southwest Kansas has the largest Q_{stable}/area value of any of the GMDs (0.070 m) but an S_Y value similar to those of the two districts to its north (0.056). However, the density of water use (0.108 m) is over twice the average of GMDs 1 and 4 to its north, and 74% greater than the average of GMDs 2 and 5 to its east. This significantly larger water-use density is likely partially responsible for the greater Q_{stable}/area through greater irrigation return flow and enhanced precipitation recharge from irrigated fields. Unlike in the other GMDs, the HPA in part of GMD3 is directly underlain by aquifer units that are often in good hydraulic connection with the HPA. Thus, pumping-induced inflow from those underlying aquifer units could also be partly responsible for the large Q_{stable}/area. In addition, the greater rate of water-level decline, coupled with the complexity of hydrogeologic conditions in GMD3, can give rise to greater amounts of drainage from recently dewatered units and units of relatively low permeability (e.g., Butler et al., 2013). The reduction required to attain stable water levels for the near-term is 33%, the largest for the Kansas GMDs;

(3) GMDs 2 and 5 – these GMDs in subhumid south-central Kansas have similar Q_{stable}/area values (0.060–0.064 m), S_Y values (0.030–0.035), and density of water use (0.061–0.063 m). The reductions required to attain stable water levels are, by far, the lowest in the Kansas GMDs (< 6%) as a result of shallower water tables, greater precipitation, and greater amount of stream-aquifer interactions. These characteristics also somewhat weaken the viability of the assumption of a constant net inflow. Changes in climate that lead to changes in the frequency of extreme precipitation events could potentially alter the long-term average net inflow to these areas. The lower S_Y values for these areas are likely produced by a combination of finer sediments resulting from the long travel distance from their source to the west and the entrapment of air and water in the vicinity of the water table produced by water-level fluctuations about a near-stable position (Butler, Whittemore, Wilson, and Bohling, 2017).

This study reveals that practically feasible pumping reductions (10–20% of average annual pumping) can have a large impact on water-level decline rates. Larger reductions (20–30%) would lead to near stable water levels, at least for the next decade or two, in the western Kansas HPA and, likely, many of the world's aquifers providing water for irrigated agriculture. The practical feasibility of various levels of pumping reductions will vary within and between aquifers. Preliminary economic analyses of the SD-6 LEMA reveal that pumping reductions have not diminished net income (Golden, 2016, 2017). Farmers in GMD4 and elsewhere in the Kansas HPA report that use of soil-moisture sensors has shown that previous irrigation strategies led to overwatering, particularly during growing seasons with average and above average precipitation (Northwest Kansas Groundwater Management District No. 4, 2017; Roth, 2017; pers. communication, Lane Letourneau, KDA, 12 September 2017). Thus, widespread adoption of soil-moisture sensors and other elements of precision agriculture could significantly enhance the prospects for pumping reductions that would have a large impact on decline rates.

Implementation of more efficient irrigation technologies, however, does not necessarily translate into actual groundwater conservation (Ward & Pulido-Velazquez, 2008). All too often, efficiency improvements appear to have led to irrigation of larger areas, double-cropping, or irrigation of more water-needy crops, and not water savings (Pfeiffer & Lin, 2014). Adoption of more efficient irrigation technologies with the goal of pumping reductions must therefore be accompanied by a means of demonstrating that actual water savings have been achieved. In addition, further work is needed to develop a better understanding of the ramifications of improvements in irrigation efficiency on the components of an aquifer's net inflow.

A major strength of the KGS-WBM approach is that it recognizes the current reality: reliable quantification of individual inflow and outflow components of an aquifer's water balance is rarely achievable. As we have shown here, lumping the inflows and outflows other than pumping together enabled us to obtain important insights into the future of groundwater-fed irrigation in a data-rich portion of the High Plains aquifer, one of the world's most important aquifers in terms of the agricultural production it supports (Marston et al., 2015). The primary assumption underlying assessments of future conditions obtained with the KGS-WBM is that the net inflow of the recent past will be the net inflow of the near future, which, given the inertia of groundwater systems, is reasonable for assessments of areas larger than a few hundred square kilometers and for time frames of one to two decades, if not longer. Assessments for durations of several decades, however, must recognize that sizable changes in net inflow may occur. Fortunately, such changes can be readily identified from ΔWL versus Q plots (Griggs & Butler, 2017).

Finally, this work has shown that much information about an aquifer's conditions and its prospects for continued support of irrigated agriculture can be obtained from data on water-level changes and water use. Although many entities have monitoring programs to assess water-level changes through time, legal, cultural, and logistical hurdles have made it difficult to obtain reliable pumping data. Over 95% of the non-domestic pumping wells in the Kansas HPA have totalizing flowmeters, all of which are subject to regulatory verification (Lanning-Rush, 2016). These water-use data, in combination with annually collected water-level data, allow us to obtain rarely available insights into an aquifer's future. Such monitoring programs should not continue to be anomalous outliers. The global water resources community must place a greater emphasis on obtaining reliable water-use data if we are to meet societal expectations and better predict prospects for extending the lifespan of aquifers that support critically needed agricultural production.

Acknowledgments

This work was supported, in part, by the Kansas Water Office (KWO), the Kansas Water Plan, and the United States Department of Agriculture (USDA). This paper was given at the workshop on Virtual Water in Agricultural Products: Quantification, Limitations and Trade Policy (Lincoln, Nebraska, USA, 14-16 September 2016), which was sponsored by the OECD Co-operative Research Programme: Biological Resource Management for Sustainable Agricultural Systems. The first author acknowledges support he received from the Nebraska Water Center to attend the workshop. Any opinions, findings, and conclusions or recommendations expressed in this material are those of the authors and do not necessarily reflect the views of the KWO, USDA, or the Nebraska Water Center.

Disclosure statement

No potential conflict of interest was reported by the author.

Funding

This work was supported by the Kansas Water Office [18-103]; Kansas Water Plan [Ogallala-High Plains Aquifer Assessment Program]; United States Department of Agriculture [Subaward RC104693B].

References

Aeschbach-Hertig, W., & Gleeson, T. (2012). Regional strategies for the accelerating global problem of groundwater depletion. *Nature Geoscience*, *5*, 853–861.

Asoka, A., Gleeson, T., Wada, Y., & Mishra, V. (2017). Relative contribution of monsoon precipitation and pumping to changes in groundwater storage in India. *Nature Geoscience*, *10*, 109–117.

Buchanan, R. C., Wilson, B. B., Buddemeier, R. R., & Butler, J. J., Jr. (2015). *The High Plains aquifer*. Kansas Geological Survey public information circular 18. Lawrence, KS: Kansas Geological Survey.

Butler, J. J., Jr., Stotler, R. L., Whittemore, D. O., & Reboulet, E. C. (2013). Interpretation of water-level changes in the High Plains aquifer in western Kansas. *Groundwater*, *51*(2), 180–190.

Butler, J. J., Jr., Whittemore, D. O., Wilson, B. B., & Bohling, G. C. (2016). A new approach for assessing the future of aquifers supporting irrigated agriculture. *Geophys Researcher Letters*, *43* (5), 2004–2010.

Butler, J. J., Jr., Whittemore, D. O., Wilson, B. B., & Bohling, G. C. (2017, December). *A major challenge for modeling conservation-based water use reductions in aquifers supporting irrigated agriculture: The specific yield quandary. Abstract H23G-1761 presented at 2017 fall meeting.* New Orleans, LA: American Geophysical Union.

Dalin, C., Wada, Y., Kastner, T., & Pum, M. J. (2017). Groundwater depletion embedded in international food trade. *Nature*, *543*, 700–704.

Golden, B. (2016). Monitoring the impacts of Sheridan county 6 local enhanced management area. Interim report for 2013–2015. Retrieved from www.agmanager.info/sites/default/files/pdf/Sheridan-LEMA_2013-2015.pdf.

Golden, B. (2017). Monitoring the impacts of Sheridan county 6 local enhanced management area, 2013–2016. Paper presented at the *2017 Governor's Conference on the Future of Water in Kansas*, Manhattan, KS. Retrieved from www.kwo.ks.gov/docs/default-source/governor's-water-conference/2017-governor's-conference-presentations/golden—impacts-of-enhanced-management-area.pdf?sfvrsn=2.

Griggs, B. W., & Butler, J. J., Jr. (2017). Groundwater in the American West: How to harness hydrogeological analysis to improve groundwater management. In P. Mulroy (Ed.), *The water problem: Climate change and water policy in the United States* (pp. 113–144). Washington, D. C.: Brookings Institution Press.

Groundwater Management District Act (2017). K.S.A. 82a-1041. Local enhanced management areas. Retrieved from agriculture.ks.gov/docs/default-source/statues-water/gmd_act.pdf?sfvrsn=12

Hanjra, M. A., & Qureshi, M. E. (2010). Global water crisis and future food security in an era of climate change. *Food Policy*, *35*(5), 365–377.

Kansas Department of Agriculture (2011). Kansas irrigation water use 2010. Retrieved from agriculture.ks.gov/docs/default-source/dwr-water-appropriation-documents/2010_irrigation_water_use.pdf?sfvrsn=2

Kansas Department of Agriculture. (2013). Order of designation approving the Sheridan 6 local enhanced management area within groundwater management district No. 4. Retrieved from dwr.kda.ks.gov/LEMAs/SD6/LEMA.SD6.OrderOfDesignation.20130417.pdf.

Kansas Department of Agriculture. (2017). Sheridan County 6 LEMA. Retrieved from agriculture.ks.gov/divisions-programs/dwr/managing-kansas-water-resources/local-enhanced-management-areas/lists/lemas/sheridan-county-6-lema.

Kansas Geological Survey. (2018). High plains aquifer atlas. Retrieved from www.kgs.ku.edu/HighPlains/HPA_Atlas/index.html

Kansas Senate Bill 156 (2015). New section 1, water conservation areas. Retrieved from www.kslegislature.org/li_2016/b2015_16/measures/documents/sb156_enrolled.pdf

Kansas Statutes Annotated 82a-1041. (2012). Local enhanced management areas; establishment procedures; duties of chief engineer; hearing; notice; orders; review. Retrieved from www.ksrevisor.org/statutes/chapters/ch82a/082a_010_0041.html.

Lanning-Rush, J. L. (2016). Irrigation water use in Kansas, 2013. *U.S. Geological Survey Data Series 981*. Reston, VA: United States Geological Survey.

Marston, L., Konar, M., Cai, X., & Troy, T. J. (2015). Virtual groundwater transfers from overexploited aquifer in the United States. *Proceedings Natural Academic Sciences of the USA, 112*(28), 8561–8566.

McGuire, V. L. (2017). Water-level and recoverable water in storage, High Plains aquifer, predevelopment to 2015 and 2013–15. *U. S. Geological Survey Scientific Investigations Report 2017–5040*. Reston, VA: United States Geological Survey.

McKee, T. B., Doesken, N. J., & Kleist, J. (1993). The relationship of drought frequency and duration to time scales. *Preprints, 8th Conference on Applied Climatology*, Anaheim, CA, 179–183.

Miller, R. D., Buchanan, R. C., & Brosius, L. (1999). Measuring water levels in Kansas. *Kansas Geological Survey Public Information Circular 12*. Lawrence, KS: Kansas Geological Survey.

National Climatic Data Center (2017). Retrieved from www7.ncdc.noaa.gov/CDO/CDODivisionalSelect.jsp#.

Northwest Kansas Groundwater Management District No. 4. (2016). Revised management program. Retrieved from gmd4.org/Management/GMD4-MgtPro.pdf.

Northwest Kansas Groundwater Management District No. 4. (2017). *The water table*. Fall 2017. Colby, KS: Northwest Kansas Groundwater Management District No. 4.

Peck, J. C. (1980). Kansas groundwater management districts. *University of Kansas Law Review, 29*, 51–91.

Pfeiffer, L., & Lin, C.-Y. C. (2014). Does efficient irrigation technology lead to reduced groundwater extraction? Empirical evidence. *Journal of Environmental Economics and Management, 67*, 189–208.

Roth, D. (2017). Water technology farms: A producer's perspective. Paper presented at the *2017 Governor's Conference on the Future of Water in Kansas*, Manhattan, KS. Retrieved from www.kwo.ks.gov/docs/default-source/governor's-water-conference/2017-governor's-conference-presentations/roth—water-technology-farm.pdf?sfvrsn=2.

Scanlon, B. R., Faunt, C. C., Longuevergne, L., Reedy, R. C., Alley, W. M., McGuire, V. L., & McMahon, P. B. (2012). Groundwater depletion and sustainability of irrigation in the US High Plains and Central Valley. *Proceedings Natural Academic Sciences of the USA, 109*(24), 9320–9325.

Smidt, S. J., Haacker, E. M. K., Kendall, A. D., Deines, J. M., Pei, L., Cotterman, K. A., … Hyndman, D. W. (2016). Complex water management in modern agriculture: Trends in the water-energy-food nexus over the High Plains aquifer. *Science of the Total Environment, 566–567*, 988–1001.

Ward, F. A., & Pulido-Velazquez, M. (2008). Water conservation in irrigation can increase water use. *Proceedings National Academy of Sciences, 105*(47), 18215–18220.

Whittemore, D. O., Butler, J. J., Jr., & Wilson, B. B. (2016). Assessing the major drivers of water-level declines: New insights into the future of heavily stressed aquifers. *Hydrological Sciences Journal, 61*(1), 134–145.

Irrigation variability and climate change affect derived distributions of simulated water recharge and nitrate leaching

Timothy R. Green and Saseendran S. Anapalli

ABSTRACT

Irrigation ('blue') water has high value as municipalities seek water security under growing populations and projected climates, but spatial variability makes estimating return flows to groundwater challenging. We demonstrate a framework for simulating spatially variable infiltration and derived distributions of return flows using an agricultural and vadose zone model to simulate recharge and nitrate leaching under irrigated corn in semi-arid northeastern Colorado, USA. Derived distributions indicated increased historical recharge (2–42%) as the spatial variability of applied irrigation increased. Projected climate in 2050 increased recharge above historical rates by up to 58%, but climatic effects decreased with increasing irrigation variability.

Introduction and motivation

Water rights, policy and emergent water markets call for improved estimates of agricultural water budgets. In Colorado, USA, water law considers historical return flows to streams and groundwater part of regional water resources which must be quantified before selling water rights. Such return flows are challenging to estimate due to spatial variability of applied irrigation water caused by inefficiency of in-field irrigation application. Irrigation field research has been conducted to evaluate crop water production functions in the semi-arid eastern plains of Colorado, USA (Saseendran, Ahuja, Nielsen, Trout, & Ma, 2008; Trout & DeJonge, 2017). Such studies are important for assessment of the water footprint and optimal use of water resources. Eastern Colorado from the Continental Divide of the Rocky Mountains to the eastern plains is a system in which water flows from pristine alpine conditions with streamflow dominated by spring snowmelt, through an urban and suburban corridor with growing water demands, to agricultural landscapes where crop production is enabled or greatly enhanced by irrigation ('blue' water consumption). Agriculture is the main water user, but cities are buying water rights for potential use upstream, because the water source is primarily coming from the mountains. Buyers (cities and industry) and sellers (agricultural producers) both need to know the value of water in this semi-arid

---

environment, and producers need to evaluate scenarios for optimizing reduced irrigation water use.

In the context of this special issue, blue water refers to irrigation water supplies from surface water and groundwater. In Colorado, surface water is typically diverted from rivers into canals ('ditches') managed by ditch companies that provide water to multiple users (Waskom, Marx, Wolfe, & Wallace 2014). Groundwater pumping is limited to regional aquifers (other than part of the Ogallala Aquifer in the southeast of Colorado). Interactions between direct precipitation ('green' water), which is highly variable, and evapotranspiration (ET) demand of a crop determine the net demand for blue water. Excess irrigation or precipitation may cause return flows to surface water and eventually to groundwater, augmenting or recharging blue water supplies. Soil water in the root zone (also called 'green water storage' by some) is actually supplied by both blue and green waters. In practice, disaggregating blue and green water can be complex and highly uncertain.

Projected climate change and variability further complicate the assessment, and require computer simulations to extend available field data and predict complex interactions in agricultural systems (e.g., Lauffenburger, Gurdak, Hobza, Woodward, & Wolf, 2018). The present study builds upon the work by Islam et al. (2012b) who simulated potential effects of climate change on corn production driven by historical (1950–1999) and projected climates centred at 2050 (2035–65) and 2080 (2065–95) using an agricultural systems and vadose zone model named the Root Zone Water Quality Model (RZWQM2) (Ma et al., 2012a). However, they did not explore irrigation effects on deep percolation of water and nitrate leaching beneath the root zone. Previous work also has not considered irrigation (in)efficiency in terms of its impact on the in-field spatial variability of irrigation, which drives subsequent variability of water and nutrient fluxes, including surface runoff (considered negligible in field research), recharge and leaching of N beneath the root zone.

Agricultural system models have been used extensively for studying the impacts of climate change on crop production. In this regard, the USDA-ARS Root Zone Water Quality Model (RZWQM2) (Ahuja, Rojas, Hanson, Shaffer, & Ma, 2000) linked with the DSSAT suite of crop models (Jones et al., 2003), has been used widely to simulate the effects of both Free Air-CO_2 eneriched Experiments (FACE) and Global Climate Model (GCM) projected temperatures and rainfall on diverse field crops and crop rotations under various rainfed (green water), irrigated (primarily blue water) and N fertilizer application levels in the Great Plains of the USA (Islam, Ahuja, Garcia, Ma, & Saseendran, 2012a, 2012b; Ko et al., 2010, 2012).

The objectives of this simulation study were to:

(1) Propose a framework for estimating the derived distributions of simulated fluxes from specified distributions of spatially variable irrigation.
(2) Quantify non-linear flux responses to variable irrigation amounts using the previously calibrated agricultural systems and vadose zone model, RZWQM2.
(3) Explore potential effects of climate change on the averages of the above derived distributions of fluxes.

Methods and study location

The framework for exploring potential impacts of irrigation efficiency and climate change is applied to continuous corn (maize; *Zea mays L.*) based on experiments at the Limited Irrigation Research Farm (LIRF) in eastern Colorado, USA (Trout & DeJonge, 2017). Islam et al. (2012a, 2012b) used RZWQM2 to simulate maize grain yield and water use efficiency assuming uniform spatial irrigation at full evapotranspiration (ET) demand and for five levels of deficit irrigation under historical and projected climates. The current simulations build upon their study using the previously calibrated model parameters (Ma et al., 2012b), as described below.

Study location: limited irrigation research farm, Colorado, USA

The LIRF field site is located near Greeley, Colorado, USA (Figure 1) just north of the confluence of the Cache la Poudre and South Platte Rivers. A detailed site description and data for maize crop water productivity are available for the years 2008–2011 (Trout & Bausch, 2017). During this period, maize was grown in a four-year rotation with winter wheat, sunflower and pinto bean, where maize was in a different block of treatment plots each year. Six irrigation treatments ranged from 40–100% of the estimated crop water requirement or full evapotranspiration (ET) demand. Experiments were designed to minimize other management treatment effects, including

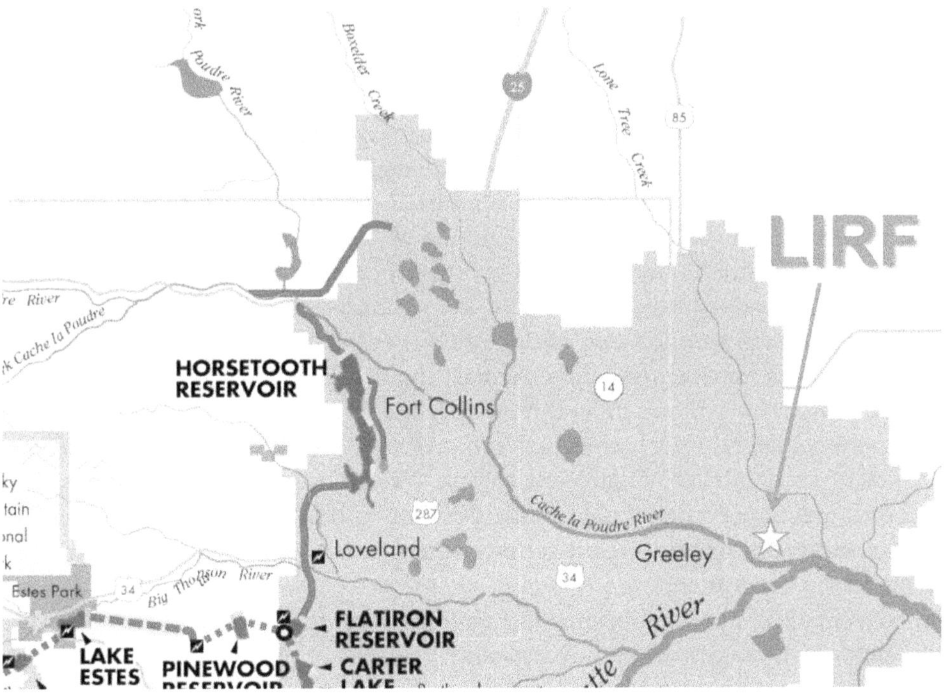

Figure 1. The Limited Irrigation Research Farm (LIRF) study site (star symbol) near Greeley, Colorado, USA. Green shaded area indicates the general extent of irrigated agriculture.

Source: Northern Water, used with permission.

seeding rate, emergence and nutrient stress. Actual conditions varied somewhat among irrigation treatments (Trout & Bausch, 2017), but effects on maize production and water use were considered minor compared with the irrigation treatment effects (Gleason et al., 2017; Trout & DeJonge, 2017).

Irrigation events were applied every 4–7 days in the field based on soil water deficits and forecasted precipitation in order to target the six treatment levels. Each treatment was replicated four times, and average seasonal irrigation amounts ranged from 126 to 427 mm of water applied. As described below, simulated irrigation events did not match the experimental rates exactly, but were designed to mimic the treatments using weekly applications based on computed ET demand. No interactions between plots (9-m wide) were assumed in all LIRF experimental and simulation results to date. Surface runoff from one plot to another is rarely observed and considered to be insignificant in the water balance. Canopy-level atmospheric conditions are likely affected by advection of air from off-site and between plots, but atmospheric variables are considered to be uniform across the LIRF experimental plots.

Newer experiments at LIRF (2012- present) were modified to optimize water used based on growth stages (unpublished data). These treatments and effects of timing allocations were not simulated here, but have been investigated in terms of maize physiological responses to water stress (DeJonge, Taghvaeian, Trout, & Comas, 2015; Gleason et al., 2017).

Agricultural systems simulations using RZWQM2

We built this study upon previous simulations (Islam et al., 2012b), where RZWQM2 was calibrated to field grain yield, biomass, leaf area index and soil water content in the root zone for three growing seasons (2008–2010). Ma et al. (2012b) documented details of the calibration procedure, which showed excellent agreement between simulated and field-estimated transpiration, as well as simulated versus experimental irrigation amounts. Evaluations of these and other simulated variables and system components provide a relatively high level of confidence in the calibrated model's responses to irrigation. Surface runoff, however, was neither measured nor tested in the model evaluations. Thus we are less confident in the simulated runoff volumes, but expect that runoff comprises a minor part of the total water balance.

Application of a vertical (one-dimensional) model like RZWQM2 may be problematic when used to explore fully three-dimensional processes. One-dimensionality assumes minimal lateral interactions between processes that would bias the representative values of each subarea. Significant surface runoff under high irrigation rates or extreme precipitation events could violate this assumption, so such events should be checked in simulation results. Again, surface runoff is not measured and is considered negligible in the experiments.

The RZWQM2 is a process-oriented cropping system model developed for simulating the impacts of tillage, residue cover, water, fertilizers, pesticides, and crop management practices on carbon, nitrogen, and water dynamics by integrating the physical, chemical and biological processes in the soil-water-crop-air system (Ahuja et al., 2000). In addition to a generic crop model that can be parameterized to simulate specific crops, the model contains the CSM (Cropping System Models) crop modules of DSSAT v4.6 for simulation of specific crops (Ma, Hoogenboom, Ahuja, Ascough, & Saseendran, 2006). Studies verifying the capabilities of RZWQM2 for explaining and managing dryland cropping systems in the Great Plains have been reported (Saseendran et al., 2008, 2015). Cameira et al. (2005, 2007) evaluated the capabilities of RZWQM to simulate crop development, water balance terms, and fate of nitrogen in field soil-crop environment. They concluded that model accuracies were acceptable in practical applications for complex and spatially variable field conditions.

The nutrient model simulates nitrogen cycling in the soil (Ma et al., 2012a). Processes include mineralization and immobilization of N based upon C:N ratios, aerobic decay of organic matter to ammonium (NH_4), aerobic nitrification of NH_4 to nitrate (NO_3), and anaerobic denitrification of NO_3 to gaseous forms (N_2 and N_2O). Soil water content, temperature, pH and microbial pools affect these processes and rates of transformation.

The model interface for RZWQM2 provides automated tools to simulate management practices including irrigation scheduling. The present simulations implemented the ET deficit method with 7 days minimum between irrigation events and 100% of the potential ET specified for the base simulation. This is defined as the 100% ET demand irrigation level, which averaged 533 mm y^{-1} for the historical period and 474 mm y^{-1} for the 2050 projected climate scenario (Islam et al., 2012b). Other levels are simply scaled by a factor (e.g., 1.4 for 140% ET Demand) applied to all irrigation events. This is appropriate for simulating a distribution of various rates on the same application dates (i.e., in-field variability).

The historical meteorological data inputs spanned 50 years (1 January 1950 – 31 December 1999) for these simulations. To remove artificial effects of initial conditions, model runs were repeated four times, saving the final state after each run and using the simulated state variables as initial conditions for the subsequent run. The model domain was 3 m deep with a unit gradient (gravity drainage) bottom boundary condition. Nutrient inputs included three forms of N applications (NO_3, NH_4 and Urea-N) applied at the experimental rates and timing each year totalling 150 kg-N ha^{-1} y^{-1}.

Daily model output included actual ET, surface runoff, recharge ('deep drainage' in RZWQM2) and N leached at 3-m depth. Here, deep percolation is considered pre-groundwater recharge without estimating a time lag, which could be substantial for a deep water table. Long-term average recharge to groundwater would be the same, but N loads could change with denitrification and other N-cycle processes above the water table.

Irrigation efficiency and derived distributions of flux variability

Irrigation efficiency within a field is related to the uniformity of irrigation application and soils, whereas whole-system irrigation efficiency also considers off-field conveyance and storage losses. Factors within a field include spatial variability of applied water (via furrows, sprinklers or drip lines) and soil infiltrability causing lateral redistribution of applied water. In these simulations, soils were assumed to be uniform in the horizontal dimensions, making irrigation uniformity the only effective factor. Field-scale irrigation efficiency or 'application efficiency' (Clemmens, Allen, & Burt, 2008) is defined as,

$$AE = \frac{100\%\ ET\ demand}{Irrigation\ water\ applied} \tag{1}$$

where 100% ET demand is the computed PET (mm d^{-1}) minus precipitation (mm d^{-1}). These fluxes and the areal average irrigation water applied (mm d^{-1}) are integrated over the growing season to get the average *AE*. On a daily basis, the change in soil-water storage should be subtracted from the denominator, but on a seasonal basis, the change in storage is assumed to be much smaller than the depth of irrigation water applied. This assumption is appropriate for well-irrigated crops, not for dryland or very limited average irrigation.

Clemmens et al. (2008) discussed various aspects of irrigation efficiency related to water conservation. They provided illustrations (Clemmens et al., 2008; Figures 2 and 3) of the relationships between the distributions of water applied and portion of area exceeding the full crop water demand for distributions with different spatial coefficients of variation. They used nonlinear, symmetric (e.g., Gaussian) distributions, but the spatial distributions could be anything from linear (uniform distribution) to highly skewed (e.g., log-normal).

If the spatial distribution of irrigation water applied is known or otherwise specified, one needs a transfer function to convert the distribution of annual irrigation water applied to a distribution of output fluxes (e.g., deep percolation). Because the transfer function $g(x)$ can be highly nonlinear, RZWQM2 was used to compute output fluxes y for discrete values of annual irrigation application amount x, which were then interpolated to obtain a continuous function for $y = g(x)$.

Figure 2 is a schematic illustration (after Benjamin & Cornell, 1970) of an output distribution or probability density function $f_Y(y)$ derived from an input distribution $f_X(x)$ and a nonlinear transfer function $y = g(x)$. The equal areas represent

$$f_Y(y)\ dy = f_X(x)\ dx \tag{2}$$

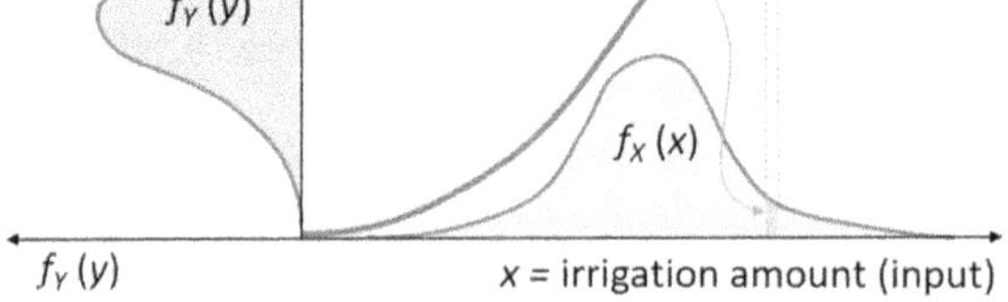

Figure 2. Conceptual diagram of a derived distribution (after Benjamin & Cornell, 1970, Fig. 2.3.4), where $f_Y(y)$ is derived from $f_X(\mathbf{x})$ given $y = g(x)$. Here, x is irrigation amount (or fraction of full ET), y is the average RZWQM2 output (deep percolation, N leached, or surface runoff), and the continuous function $g(x)$ is interpolated from discrete simulations (see results below in Figure 5).

The schematic distributions are bell shaped, but here we assume that $f_X(x)$ follows a simple uniform distribution, meaning $f_X(x) = (x_{max} - x_{min})^{-1}$ for all x within this finite range, and the mean irrigation water applied is $(x_{max} + x_{min})/2$. Also, we let $g(x)$ be a polynomial function, because this continuous analytical function fit the discrete model results very well. The derived distribution approach would work for any specified input distribution (enforcing $x \geq 0$) and any interpolated transfer function $y \sim g(x)$ fitted to the RZWQM2 output fluxes.

In the present simulations, the mean value of irrigation water applied meets 1.4 times the computed ET demand, and the range is allowed to vary (Table 1). This assumption is based on common irrigation practices that avoid under-watering large parts of a field, leading to higher field-average irrigation amounts. This differs from the field experiment at LIRF, which avoided over-irrigation and targeted deficit irrigation rates as low as 40% of full ET demand. In our simulations, irrigation rates varied from 30% to 250% of full ET demand for the highest variability case, such that $0.3 \leq x \leq 2.5$ in the interpolated the transfer function $g(x)$.

Finally, the spatial mean value of each output flux is the integral over the derived distribution, or

$$\bar{y} = \int_{-\infty}^{\infty} f_Y(y)dy \tag{3}$$

where again, y represents either recharge, N leaching, or surface runoff integrated over the full time period of simulation. In this application the lower limit of integration is 0, because fluxes are always non-negative. In general, the lower limit could be negative.

Downscaled ensemble of projected climate change

The projected climate scenarios used for this study were taken from Islam et al. (2012b), who used the Hybrid Delta Method (Hamlet, Salathé, & Carrasco, 2010) with bias corrected and spatially disaggregated (BCSD) climate projections from the World Climate Research Program's (WCRP's) Coupled Model Intercomparison Project phase 3 (CMIP3) dataset (IPCC, 2007). Details of the climate downscaling can be found in Islam et al. (2012b). The CMIP3 projections used 16 general circulation

Table 1. Derived average fluxes under historical climate (1950–1999) for different ranges of irrigation amounts, and corresponding ratios of these averages divided by fluxes for uniform irrigation at 140% of full ET demand.

Irrigation Fraction of Full ET Demand			Derived Average Flux			Ratio vs 140%		
Range	Minimum	Maximum	Recharge (mm/y)	N leached (kg/ha/y)	Runoff (mm/y)	Recharge	N Leached	Runoff
0.4	1.2	1.6	115	4.23	8.78	1.02	1.03	0.999
0.8	1.0	1.8	120	4.30	8.78	1.07	1.05	0.998
1.0	0.9	1.9	124	4.36	8.78	1.10	1.07	0.998
1.4	0.7	2.1	134	4.52	8.78	1.19	1.11	0.998
1.8	0.5	2.3	153	4.93	8.77	1.36	1.21	0.997
2.2	0.3	2.5	160	5.12	8.77	1.42	1.25	0.997

models (GCMs) with three greenhouse gas emission scenarios representing low (B1), medium (A1B) and high (A2) emissions paths. The simulated ensemble average atmospheric CO_2 concentrations were 330 ppm for the historical period, and 460, 623, and 790 ppm for projections centered on the years 2020, 2050, and 2080, respectively.

CMIP3 climate projections are used for consistency with previous simulations, noting that the current emphasis is on the framework for simulating effects of irrigation efficiency. One may question the use of CMIP3 versus more recent CMIP5 climate simulations (IPCC, 2013), because more climate models were used for CMIP5, and some of the newer GCMs are considered more advanced. CMIP5 used Representative Concentration Pathways (RCPs), which differ from the emissions scenarios used in CMIP3, making direct comparisons difficult. However, projections from CMIP3 and CMIP5 are generally consistent.[1] For North America, Sheffield et al. (2014) concluded, 'Overall, the multi-model ensemble (MME) mean performance has not improved substantially in CMIP5 relative to CMIP3 for climatological variables … Projected increases in moderate to extreme precipitation events are similar to CMIP3.' Sun et al. (2015) provided additional details on the similarities and differences between CMIP3 and CMIP5 climate projections for the United States. Soon CMIP5 will be superseded by CMIP6 (Eyring et al., 2016) and so on, which is why we focus here on the methods for simulating responses to irrigation variability more than on the specific climate change results.

Results

Example results are presented for the 50-year historical climate and for one climate projection period (50 years of simulation centered on the year 2050). We also ran equivalent simulations for near-term (2020) and longer-term (2080) climate scenarios, yielding more subdued and amplified results, respectively, but the illustrations would be similar to the examples shown below. As noted in the methods above, each period was simulated four times, wherein the final simulation state of each run was saved and used to reinitialize the next run, thereby reducing artificial effects of the initial conditions (particularly soil moisture and nutrient pools) on the final results.

Simulated fluxes

Historical climate data (1950–1999) include inter-annual and intra-annual variability. In particular, precipitation amounts within the growing season and extreme events vary from year to year. Daily historical precipitation amounts are shown together with simulated evapotranspiration (ET) in Figure 3(a). Despite variability in precipitation, ET is relatively consistent from year to year under full irrigation (100% of ET demand). On the other hand, recharge of water beneath the root zone ('deep percolation' at 3 m) varies dramatically among the 50 years simulated, and Figure 3(b) shows four large recharge events with peak fluxes exceeding 2 mm/d (indicated with green circles). Likewise, nitrate leaching follows the temporal pattern of recharge. Daily surface runoff is also simulated with several peaks within a range of 2–6 mm/d. However, these events

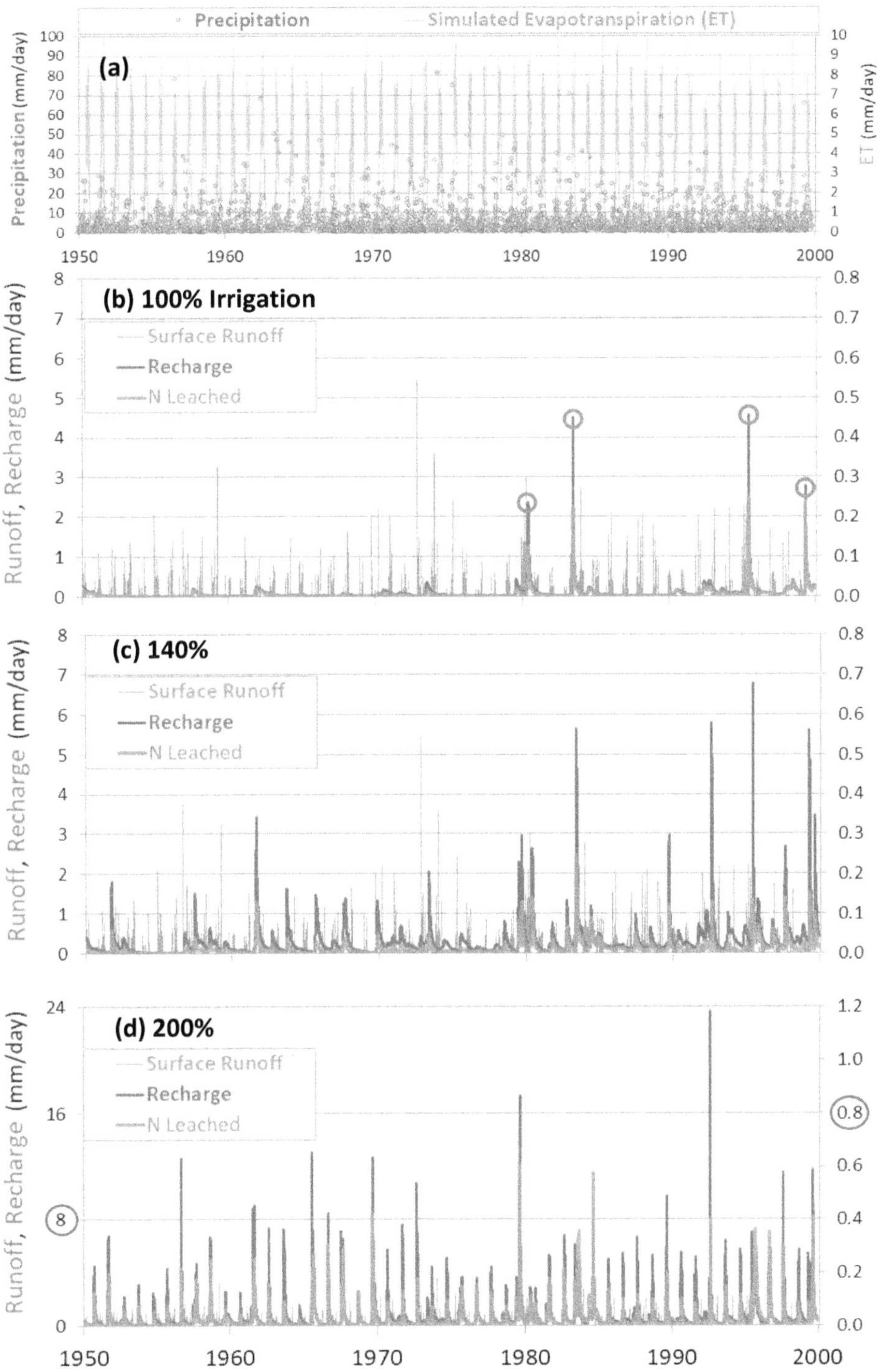

Figure 3. Daily simulation results for historical climate (1950–1999): (a) Precipitation data and simulated evapotranspiration (actual ET); and simulated runoff, recharge and N leached under irrigation meeting: (b) full (100%), (c) 140% and (d) double (200%) of the full ET demand. Green circles highlight four main recharge events with peaks >2 mm/d in (b). In (d), the red circles highlight the maximum scales used in (b) and (c).

are infrequent, such that annual runoff volumes are very small (<8.7 mm y^{-1}) compared with precipitation (average annual equals 341 mm y^{-1}).

The base scenario used here for uniform irrigation is 140% of full ET demand (Figure 3(c)). Although simulated ET increased slightly compared with the scenario of irrigation at 100% of ET demand (Figure 3(b)), ET under full irrigation is driven primarily by the energy budget, so in both cases, ET approximately met the potential ET demand. Thus excess irrigation water is available to increase soil-water storage and eventually flow below the root zone. Consequently higher irrigation produced more frequent recharge events (Figure 3(c), with peaks approaching or exceeding 1 mm/d in many years, and significantly higher recharge in the four years with high peaks under irrigation at 100% ET demand (see green circles in Figure 3(b)). Again N leaching follows the temporal pattern of recharge, so much more nitrate is leached through the profile to the 3-m depth.

Finally, Figure 3(d) shows even greater output fluxes when irrigation is applied to meet 200% of ET demand for the historical climate. In this case, significant water recharge and N leaching events occur every year, and there are 8 recharge peaks exceeding 10 mm/d (note the greater maximum value of the vertical axis compared with Figure 3(b,c)). However, simulated ET continues to meet the potential ET (energetic demand), and surface runoff does not increase with increased irrigation.

Figure 4 is a one-year window (1980) showing a zoomed in view of Figure 3 that includes the lowest of the top four recharge events (first green circle in Figure 3(b)) over the 50-year historical period. This illustrates dynamics within the growing seasons and effects of Spring precipitation. Thus, the previously calibrated RZWQM2 model captures the agricultural system and vadose zone processes needed to develop transfer functions from irrigation inputs under different periods to output fluxes, including water recharge and nitrate leaching at 3 m and surface runoff.

Results for the ensemble climate change scenario are not illustrated here, but the temporal behaviours are similar to the historical simulations. An important difference in the simulated transpiration process is the plant response to increased atmospheric CO_2. In the plant model (DSSAT 4.6), the radiation use efficiency increases with CO_2 to produce higher average leaf areas, and stomatal conductance decreases to conserve water. In combination, the average ET is reduced, which causes the average recharge under climate change (2050 scenario) to be greater than historical recharge.

Transfer functions: irrigation inputs to model output fluxes

As described in the methods, average annual input (irrigation) and output (recharge, N leaching, and runoff) are computed for each irrigation scenario (spatial representation). These simulated data are then fit using polynomial regression to compute a relationship of the form:

$$y = g(x) = ax^3 + bx^2 + cx + d \tag{4}$$

where the symbols a through d are fitting parameters, and selection of a 3rd order polynomial is based upon achieving nearly perfect fits (R^2 > 0.99) for the recharge and

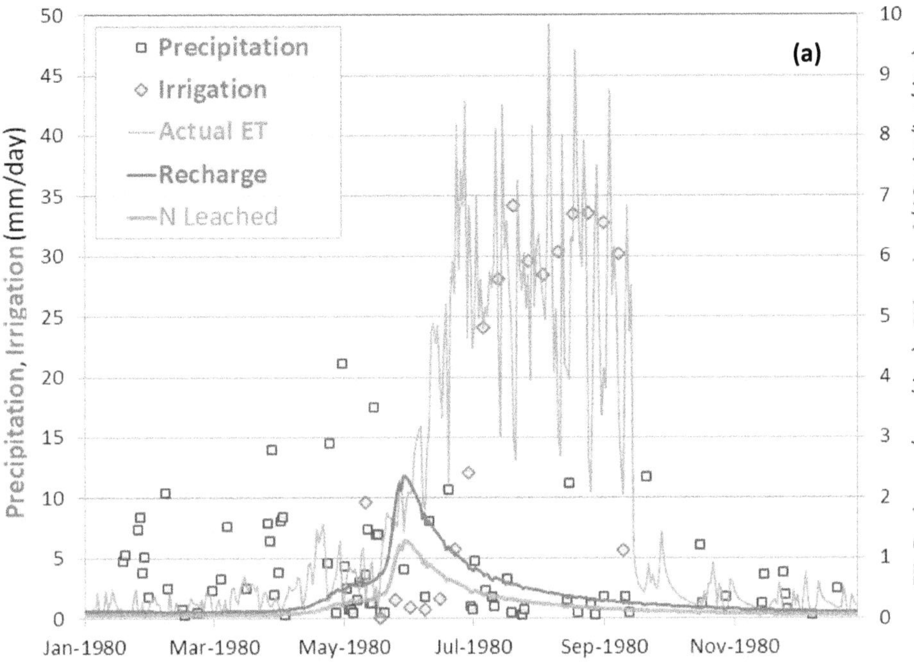

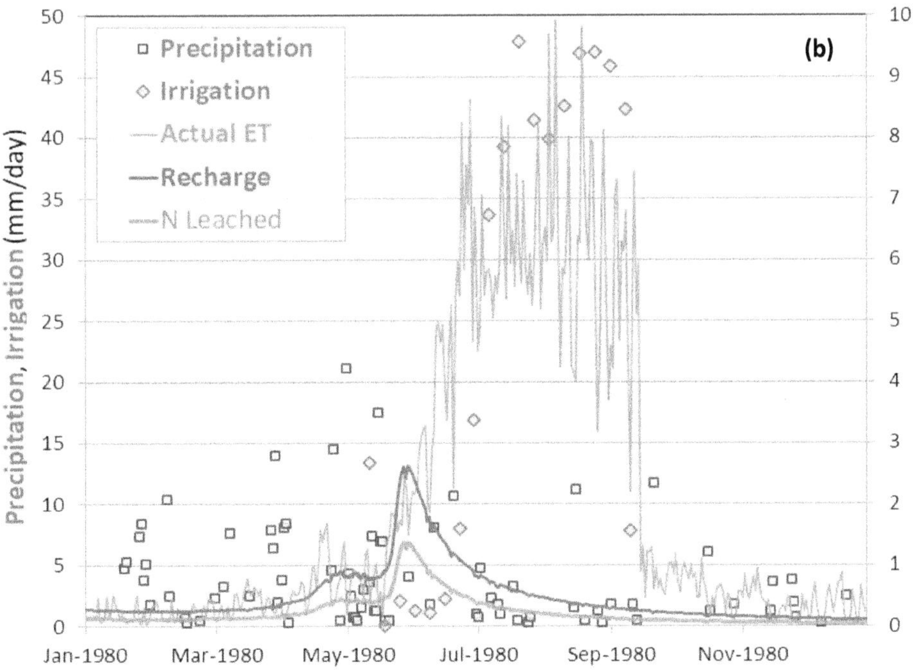

Figure 4. Daily simulation results illustrating one year (1980) under historical climate using irrigation schedules of (a) 100% and (b) 140% of the full ET demand (zoom-in view of Figure 3).

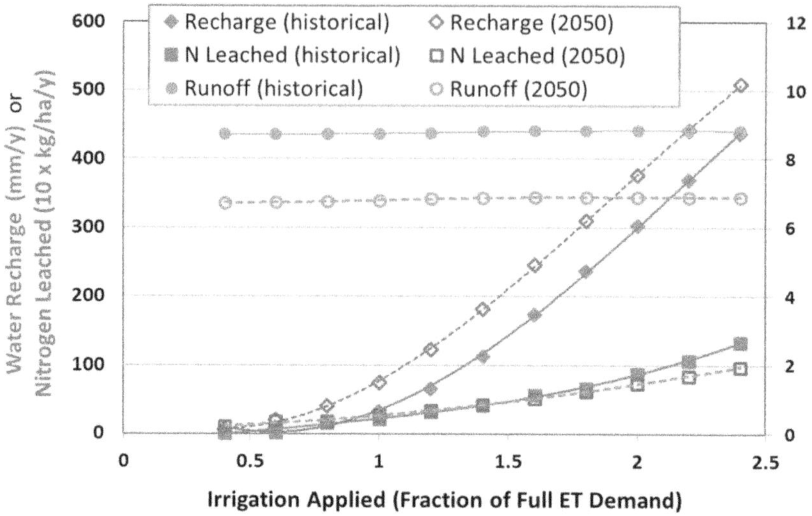

Figure 5. Average values of simulated deep percolation and nitrogen (NO_3-N) leached to 3-m depth as functions of the fraction of full ET demand for historical climate and projected climate (2050). Symbols are computed from RZWQM2 simulations for surface water runoff (green), deep percolation of water (blue) and nitrogen (NO_3-N) leached (red); lines are polynomial regressions through these points used as transfer functions (*g*(*x*) in Figure 2).

N leaching transfer functions. Figure 5 shows the resulting transfer functions for historical and 2050 projected climate scenarios.

Recharge transfer functions behave as expected with nonlinear responses to irrigation between 0.4 and 1.5 times full ET demand, followed by nearly linear increases at higher irrigation rates. The nonlinear feature requires a system model like RZWQM2 to capture complex biophysical interactions over time. Interestingly, the nonlinear portion goes beyond irrigation amounts designed to meet 100% of full ET, so the inflection point of linear extension is also identified by these simulations. Simulated nitrate leaching at 3 m is also nonlinear, but the monotonic curve never becomes completely linear even at irrigation rates exceeding 2 times full ET.

Comparing the 2050 projected climate with historical climate results, recharge increased consistently for all irrigation rates. On the other hand, simulated surface runoff decreased, but absolute changes were very small (~ 2 mm y^{-1}). Responses of N leaching to climate change varied with the irrigation scheme, showing increased N leaching under deficit irrigation (<100% of full ET demand), but decreased at levels above approximately 1.5 times full ET demand for irrigation. The projected change in N leaching is thus very small for uniform irrigation at the base rate of 1.4 times full ET demand.

Derived distributions and average fluxes

For simplicity, the input distribution for irrigation variability was assumed to be a uniform distribution for the present results. The derived distributions were computed using the polynomial transfer functions (Equation 4) in Figure 5, which were averaged by discrete integration of Equation (3). The derived average fluxes are tabulated in

Table 2. Differences (2050 Projection − Historical) for deep percolation (recharge at 3-m depth), N leached, and surface runoff. Mean irrigation applied is 1.4 times the full evapotranspiration (ET) demand for all cases with uniform distributions for the given ranges from minimum (Min.) to maximum (Max.) levels. Average flux differences and percentage changes between projected and historical climates are computed from derived distributions using simulated transfer functions (Figure 5).

Irrigation Fraction of Full ET Demand			Derived Average Flux Difference			Percentage change (%)		
Range	Min.	Max.	Recharge (mm/y)	N leached (kg/ha/y)	Runoff (mm/y)	Recharge	N Leached	Runoff
0.4	1.2	1.6	66.7	−0.046	−1.93	57.9	−1.1	−21.9
0.8	1.0	1.8	64.4	−0.061	−1.93	53.4	−1.4	−22.0
1.0	0.9	1.9	62.8	−0.076	−1.93	50.5	−1.7	−22.0
1.4	0.7	2.1	59.0	−0.132	−1.94	43.9	−2.9	−22.1
1.8	0.5	2.3	55.1	−0.242	−1.94	37.6	−5.1	−22.1
2.2	0.3	2.5	51.8	−0.435	−1.94	32.4	−8.5	−22.2

Table 1 for the historical climate with different levels of spatial variability in irrigation quantified by the range of irrigation amounts quantified by the fraction of full ET demand. Water recharge increased from 115 to 160 mm y^{-1} as the spatial variability (range) increased from 0.4 to 2.2. The average flux of N leached to a 3-m depth also increased with spatial variability of irrigation, but only from 4.23 to 5.12 kg-N ha^{-1} y^{-1}. The derived average surface water runoff changed very little, decreasing from 8.78 to 8.77 mm y^{-1} (note that runoff is less than 8% of recharge, and the change in runoff is insignificant).

The derived average fluxes with spatially variable irrigation are compared with uniform irrigation at the same mean value (140% of full ET demand) on the right side of Table 1. As expected, the lowest variance (small range) produced average values very similar to the uniform irrigation case (ratio = 1). In the most extreme spatial variability explored, the average recharge increased by 42% (ratio = 1.42), and N leached increased by 25%. Runoff did not increase with greater irrigation variability, indicating that irrigation events did not produce runoff in these simulations.

Similar tables could be given for the climate change scenario projected into the short and long term (2020 to 2080). Table 2 reports the differences between derived average fluxes using projected climate for 2050 versus the historical simulations (Table 1). Differences in recharge are positive (i.e., simulated recharge in 2050 is greater than historical values), but the differences decrease as the range of spatial irrigation values increase. The corresponding percentage change went from 58% down to 32%. By contrast, changes in N leached and runoff were all negative but small in absolute terms. The absolute change in N leached was less than 0.44 kg-N ha^{-1} y^{-1}, and runoff decreased by less than 2 mm y^{-1}.

Discussion of results

The present study is primarily an illustration of the framework presented for derived flux distributions and the associated ensemble average effects of irrigation efficiency under historical and projected climate change effects on deep water percolation (pre-groundwater recharge) and N leaching. Surface runoff was also simulated, but found to be minimal even under excessive (>200% of ET demand) irrigation. A caveat of the latter result is that RZWQM2 was not calibrated to match observed runoff from the field plots, because runoff was not measured and assumed to be minimal in field research. Surface detention storage is not simulated in RZWQM2, which would further reduce simulated runoff and even eliminate many of the small events (e.g., <2 mm d^{-1}). It is possible that careful study of runoff could yield different model results, but that is beyond the current scope. Thus surface runoff results are reported for the calibrated model, but not emphasized here.

The vertical fluxes of water and N below the root zone at 3-m deep showed dramatic temporal variability. As illustrated in Figure 3, simulation over a long time period (50 years in this study) is essential for capturing episodal recharge and leaching events under full (100% of ET demand) irrigation. Even at 140% of full ET, interannual variability was pronounced.

The resulting transfer functions were nonlinear and could not be estimated without an agricultural systems and vadose zone process model, which simulated interactions of highly nonlinear processes. Because the runoff component was not calibrated against measurements, we hesitate to draw any firm conclusions from those results, and more work is needed to investigate the transfer functions for runoff. Transfer functions for water recharge and N leaching were similar for both historical and projected climates. The simulated effects of climate change caused recharge under high irrigation to be greater in the projected climate, but the amount of N leached decreased. For now, we hypothesize that the nitrogen cycling processes in the model caused greater denitrification in the root/vadose zone due to higher profile soil-water contents and higher surface temperatures. Further study is needed to explore this in more detail.

Conclusions

The present study led to the following conclusions:

(1) A systems model such as RZWQM2 is needed to simulate deep percolation of water and N leaching under historical and projected climates.
(2) Temporal averaging of fluxes over a sufficient time period can provide well-defined transfer functions, as illustrated here, to convert variable inputs (irrigation) to the output fluxes of interest.
(3) Irrigation 'return flows' to groundwater (renewable 'blue' water) respond nonlinearly to irrigation (in)efficiency.
(4) The derived distribution approach allows distributional uncertainty of irrigation efficiency (spatial variability) to be translated to output distributions and mean fluxes over space and time.
(5) The current simulations of climate change showed potential for large increases (32 to 58%) in recharge (2050 versus historical climate).
(6) Smaller negative changes in nitrate leaching indicated complex N cycling processes may have a compensating effect on increased recharge fluxes under climate change, despite monotonic increases in nitrate leaching with increased irrigation under both historical and projected climates.
(7) The water footprint of cropping systems, including blue and green water, can be estimated along with the combined return flow to groundwater available for

future pumping (more blue water). Such analyses may require complex systems models and should not be oversimplified.

Note

1. http://climatechange.environment.nsw.gov.au/Climate-projections-for-NSW/About-NARCliM/CMIP3-vs-CMIP5.

Acknowledgments

The simulations here used climate scenarios and calibrated RZWQM2 parameters from Drs. Adlul Islam and Liwang Ma, who worked with Dr. Saseendran. Dr. Laj Ahuja was the Research Leader of the former ARS Agricultural Systems Research Unit, and his support as supervisor of both authors at that time is greatly appreciated. Original climate and cropping systems data and management used in the model were provided by the former ARS Water Management Research Unit in collaboration with Drs. Tom Trout and Kendall DeJonge. This paper was given at the workshop on Virtual Water in Agricultural Products: Quantification, Limitations and Trade Policy (Lincoln, Nebraska, USA, 14-16 September 2016), which was sponsored by the OECD Co-operative Research Programme: Biological Resource Management for Sustainable Agricultural Systems whose financial support made it possible for the author to participate in the workshop.

Disclosure statement

No potential conflict of interest was reported by the authors.

ORCID

Timothy R. Green http://orcid.org/0000-0002-1441-8008
Saseendran S. Anapalli http://orcid.org/0000-0001-5238-1894

References

Ahuja, L. R., Rojas, K. W., Hanson, J. D., Shaffer, M. D., & Ma, L. (editor). (2000). *Root zone water quality model: modeling management effects on water quality and crop production*. Highland Ranch, CO: Water Resources Publ., LLC.

Benjamin, J. R., & Cornell, C. A. (1970). *Probablility, statistics, and decision for civil engineers*. New York: McGraw-Hill.

Cameira, M. R., Fernando, R. M., Ahuja, L. R., & Ma, L. (2007). Using RZWQM to simulate the fate of nitrogen in field soilcrop environment in the Mediterranean region. *Agricultural Water Mgmt*, *90*, 121–136.

Cameira, M. R., Fernando, R. M., Ahuja, L. R., & Pereira, L. S. (2005). Simulating the fate of water in field soil-crop environment. *Journal of Hydrology*, *315*, 1–24.

Clemmens, A. J., Allen, R. G., & Burt, C. M. (2008). Technical concepts related to conservation of irrigation and rainwater in agricultural systems. *Water Resource Research*, *44*, W00E03.

DeJonge, K. C., Taghvaeian, S., Trout, T. J., & Comas, L. H. (2015). Comparison of canopy temperature-based water stress indices for maize. *Agricultural Water Manage*, *156*, 51–62.

Eyring, V., Bony, S., Meehl, G. A., Senior, C. A., Stevens, B., Stouffer, R. J., et al. (2016). Overview of the Coupled Model Intercomparison Project Phase 6 (CMIP6) experimental design and organization. *Geoscience Model Developments*, *9*, 1937–1958.

Gleason, S. M., Wiggans, D. R., Bliss, C. A., Comas, L. H., Cooper, M., DeJonge, K. C., … Zhang, H. (2017). Coordinated decline in photosynthesis and hydraulic conductance during drought stress in Zea mays. *Flora, 227*, 1–9.

Hamlet, A. F., Salathé, E. P., & Carrasco, P. (2010). *Statistical downscaling techniques for global climate model simulations of temperature and precipitation with application to water resources planning studies* (Technical report). Retrieved from https://digital.lib.washington.edu/researchworks/bitstream/handle/1773/38428/2010-5.pdf?sequence=1

IPCC. (2007). *Climate change 2007: The physical science basis. Contribution of Working Group I to the Fourth Assessment report of the intergovernmental panel on climate change* (p. 996). Cambridge,UK, and New York, USA: Cambridge University Press.

IPCC. 2013. Climate Change 2013: The Physical Science Basis. Contribution of Working Group I to the Fifth Assessment Report of the Intergovernmental Panel on Climate Change. Cambridge, United Kingdom and New York, NY, USA: Cambridge University Press.

Islam, A., Ahuja, L. R., Garcia, L.A., Ma, L., & Saseendran, S.A. (2012a). Modeling the effect of elevated CO2 and climate change on reference evapotranspiration in the semi-arid great plains. *Transactions of the Amererican Society of Agricultural and Biological Engineers, 55*(6), 2135-2146.

Islam, A., Ahuja, L. R., Garcia, L. A., Ma, L., Saseendran, A. S., & Trout, T. J. (2012b). Modeling the impacts of climate change on irrigated corn production in the Central Great Plains. *Agricultural Water Manage, 110*, 94–108.

Jones, J.W., Hoogenboom, G., Porter, C.H., Boote, K.J., Batchelor, W.D., Hunt, L.A., & Ritchie, J. T. (2003). The DSSAT cropping system model. *European Journal Of Agronomy, 18*, 235-265.

Ko, J., Ahuja, L.R., Kimball, B., Saseendran, S.A., Ma, L., Green, T.R., Ruane, A.C., Wall, G.W., Pinter, P., & Bader, D.A. (2010). Simulation of free air CO2 enriched wheat growth and interactions with water, nitrogen, and temperature. *Agricultural & Forest Meteorology, 150*, 1331–1346. doi:10.1016/j.agrformet.2010.06.004

Ko, J, Ahuja, L.R., Kimball, B.C., Saseendran, S.A., Ma, L., Green, T.R., Wall, G., & Pinter, P. (2012). Simulation of climate change impacts on cropping systems in the central great plains. *International Journal of Climatic Change, 111*,445–472.

Lauffenburger, Z. H., Gurdak, J. J., Hobza, C., Woodward, D., & Wolf, C. (2018). Irrigated agriculture and future climate change effects on groundwater recharge, northern High Plains aquifer, USA. *Agricultural Water Manage, 204*, 69–80.

Ma, L., Ahuja, L. R., Nolan, B. T., Malone, R. W., Trout, T. J., & Qi, Z. (2012a). Root Zone Water Quality Model (RZWQM2): Model use, calibration, and validation. *Transactions of the ASABE, 55*(4), 1425–1446.

Ma, L., Hoogenboom, G., Ahuja, L. R., Ascough, J. C. I., & Saseendran, S. A. (2006). Evaluation of the RZWQM-CERES-Maize hybrid model for maize production. *Agricultural Systems, 87*, 274–295.

Ma, L., Trout, T. J., Ahuja, L. R., Bausch, W. C., Saseendran, S. A., Malone, R. W., & Nielsen, D. C. (2012b). Calibrating RZWQM2 model for maize responses to deficit irrigation. *Agricultural Water Manage, 103*, 140–149.

Saseendran, S. A., Ahuja, L. R., Nielsen, D. C., Trout, T. J., & Ma, L. (2008). Use of crop simulation models to evaluate limited irrigation management options for corn in a semiarid environment. *Water Resources Researcher*, (44), W00E02. doi:10.1029/2007WR006181

Saseendran, S. A., Trout, T. J., Ahuja, L. R., Ma, L., McMaster, G. S., Andales, A. A., Chávez, J. L., Ham, J. (2015). Quantification of crop water stress factors from soil water measurements in limited irrigation experiments. *Agricultural Systems, 137*, 191–205.

Sheffield, J., Barrett, A., Barrie, D., Camargo, S. J., Chang, E. K. M., Colle, B., ...Zhao, M. 2014. Regional climate processes and projections for North America: CMIP3/CMIP5 differences, attribution and outstanding issues. NOAA Technical Report OAR CPO-2. MD: Silver Spring.

Sun, L., Kunkel, K. E., Stevens, L. E., Buddenberg, A., Dobson, J. G., & Easterling, D. R. 2015. Regional surface climate conditions in CMIP3 and CMIP5 for the United States: Differences, similarities, and implications for the U.S. National climate assessment. NOAA Technical Report, NESDIS 144. p. 111.

Trout, T. J., & Bausch, W. C. (2017). USDA-ARS Colorado maize water productivity data set. *Irrigation Sciences*, *35*, 241–249.

Trout, T. J., & DeJonge, K. C. (2017). Water productivity of maize in the US high plains. *Irrigation Sciences*, *35*, 251–266.

Waskom, R., Marx, E., Wolfe, D., & Wallace, G. (2014). *Irrigation ditches and their operation Colorado State University Extension, Natural Resources Series on Water* (Fact Sheet No. 6.701). Retrieved from http://extension.colostate.edu/topic-areas/natural-resources/irrigation-ditches-and-their-operation-6-701/

The water footprint challenge for water resources management in Chilean arid zones

Pablo Álvarez

ABSTRACT

The water footprint concept is difficult to apply to water-shortage conditions because of private management and variability in the water balance. Chilean water law stipulates that water allocation depends on water rights and private organization criteria, according to the basin water supply and not necessarily crop irrigation demand, crop yield or irrigation technology. This paper explores the resultant complications.

Introduction

The introduction of the water footprint (WF) as a concept provides a new perspective on water resource usage and management analysis in crop production. The key to guarantee agriculture's permanent development is to balance when water resources are used, adopting the right irrigation conditions for continuous drought (Qin, Jin, Duan, & He, 2016).

Agricultural is the biggest consumptive water user around the world and it has a great impact in the global water balance. Many studies on the WF and virtual water (VW), that is, 'invisible' water embedded in the life cycle of a product (Allan, 1998), are centred on agricultural commodities, especially basic food crops, because of the high water consumption for production and the importance of these in food security (Yang, Pfister, & Bhaduri, 2013). The specific-crop VW content is a function of weather conditions, agronomical practices and field management. The value varies according to the geographical location and scale of the available data (Yang & Zehnder, 2007).

At a global level, 86.5% of water consumed during crop production is green water, including precipitation and soil moisture (Mekonnen & Hoekstra, 2011). Even in irrigated agriculture, green water often makes a very significant contribution to the total water consumption. The blue water footprint (BWF) from surface water and groundwater is bigger in arid and semi-arid regions such as the west coast of South America (Peru–Chile). Irrigated agriculture in arid zones is supposed to supply enough water to satisfy the crops' water needs. However, farmers can decide whether to provide irrigation water under optimal performance. In arid zones, the assumption of adequate irrigation could imply an overestimation of BWF (Mekonnen & Hoekstra, 2011).

The WF links the amount of water used in the production stage to the finished product, typically measured in kilograms. According to Pellicer, Perni, and Martinez (2013) and Yang and Zehnder (2007), the WF is related to resource efficiency and sustainability in a production cycle. The concept of WF can be useful in areas where there is a shortage of water relative to demand. The Coquimbo Region, located in the Chilean arid zone, is one of these areas. In this region, the precipitation represents 10–20% of the reference evapotranspiration (ET_0), and it is concentrated in winter, outside the commercial growing-crop period, leading to a water deficit that compels irrigation. Therefore, the BWF is greater than the green WF for agricultural products in this region.

This paper seeks to identify the constraints and challenges to including WF in water resources management in a Chilean arid zone represented by the Coquimbo Region. The discussion considers the approach of the former concepts in an arid zone facing a legal framework that promotes the private administration of water and water rights.

The paper is structured as follows. It begins with a general description of the arid conditions and legal context related to water rights and private water management in basins of the Coquimbo Region. It then presents a discussion about the contribution of WF to water resources management in the Coquimbo Region. The paper concludes with some lessons and recommendations to improve water resources management, including the BWF and blue water balance.

Aridity, the Chilean Water Code and the private administration of water resources in the Coquimbo Region

The Coquimbo Region extends from 29° 02′S to 32° 16′S latitude. It is bordered to the west by the Pacific Ocean and to the east by Argentina. It is bordered to the north and south by the Atacama and Valparaiso regions respectively. Its total area is 40,579.9 km^2 (Universidad de Los Lagos, 2009).

This region is characterized by an arid Mediterranean climate: low precipitation and large variation from year to year. It experiences a dry period for 9–11 months, and winter precipitation from June to August (Corporación Nacional del Medio Ambiente (CONAMA), 2002).

Precipitation is concentrated during the winter and ranges, on average, from 100 to 200 mm/year, being higher in the central–southern zone of the region and lower to the north of the region (Figure 1). It is possible to observe drought periods of variable duration (2–5 years) across the Coquimbo Region, including extremely dry years (1920, 1960, 1962, 1964, 1969, 1979, 1988, 1990, 1995, 1998 and 2014) with less than 60 mm annual precipitation (CONAMA, 2002). This region has three main basins: Elqui, Limarí and Choapa. Streams in these basins have their source in the Andes Mountains and they flow to the Pacific. The Andes area, over 1500 masl, represents about the 25% of each basin surface; however, it provides more than 70% of the water flowing in the rivers (Dirección General de Aguas (DGA), 2004a, 2004b, 2004c). The total irrigated area, including the intakes of canals, represents around 3% of the three main basins' area (Figure 1).

In Chile, the Water Code has been in force since 1981 and it represents the legal framework on water resources. According to the code, permanent water rights allow

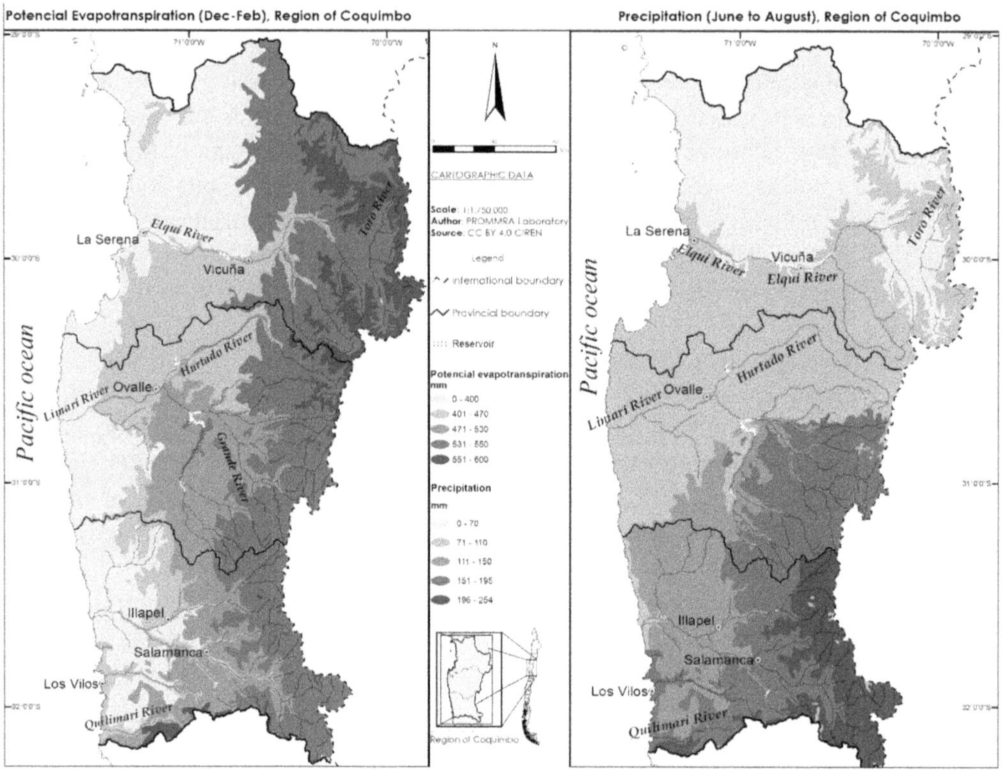

Figure 1. Precipitation distribution and potential evapotranspiration in the Coquimbo Region.
Source: Adapted from CIREN (1997).

water usage permanently depending on the supply source (Díaz, 2008). The code establishes in the 5th article that: 'Water is a national good for public use'; and in the 6th article establishes that the right to use water is given to private individuals who are able to use, benefit and arrange these rights.

A 'right' is a proportion of water available in the water source, such as a river, canal or reservoir. According to Bauer (1997), the main concerns in the code are:

- The strengthening of the private property condition for the water rights.
- Distinguishing water rights from land rights.
- The self-determination and the change of the water destination according to the owner's will, meaning there are no use priorities.
- Establishing that the rights are provided by the state, free of cost and in perpetuity.
- The state has low interference in solving disputes.
- Irrigation (agriculture) is the most important matter for this code.

According to the law, the rights are only given in the water supply sources that have not been declared depleted by the authority. In the case of Limarí Basin, there has been a depletion decree since 2005. Therefore, the only way to obtain permanent water rights is in the market (Díaz, 2008).

The parties involved in water management include the public; state and private organizations; and irrigation farmers who are the water right owners (individuals or associated). In the Chilean private sector, the owners of water rights who share the same source of water can create a non-profit organization: Organizaciones de Usuarios de Agua (OUAs). OUAs distribute water among the water right owners (shareholders). Some shareholders are permitted to be part of the OUAs directory if they are elected according to a voting process based on water rights (one water right = one vote). The Water Code promotes private organizations and makes them autonomous, under the consideration that water management related to distribution depends on directory decision-making (Díaz, 2008).

Water user organizations are classified into: (1) 'Juntas de Vigilancia', which in general are responsible for distributing water from rivers to canals; and (2) 'Comunidades de Agua', which are in charge of water distribution from irrigation canals to fields.

In the Coquimbo Region (Figure 2), the surface area irrigated with 85% probability is distributed nearly evenly among the three regional basins. A total of 39.4% is located in Limarí Basin, 30.9% in Elqui Basin and 29.7% in Choapa Basin.

A total of 102,267 water rights are registered and recognized by the national water bureau in the region: 57.3% belong to Limarí Basin, 27.7% to Elqui Basin and 15% to Choapa Basin (Díaz, 2008).

The surface water from Elqui and Choapa river basins are administrated by the Junta de Vigilancia of the Elqui and Choapa rivers respectively (Zavala & Trigos, 2008). Limarí Basin has three reservoirs (La Paloma, 0.75 km^3; Cogotí, 0.15 km^3; Recoleta, 0.1 km^3) and nine private organizations that distribute the water. They form 'La Paloma' system. The system uses the volume stocked in the three reservoirs according to an operation rule that establishes the water available for the irrigation season (May–April) (Vivanco, 2014). The operation rule establishes the water available for each organization. Each organization gathers the private members' water rights and, thus, the water allocation per water right does not consider the private demand or the kind of crop.

The Choapa River basin includes the Corrales (0.05 km^3) and El Bato (0.026 km^3) reservoirs and distributes the water according to detachment (Comisión Nacional de Riego (CNR), 2004). Detachment, just like in Elqui Basin, corresponds to a flow in l/s, which is a percentage of a water right that is distributed during the irrigation season (September–August) according to the river flow forecast and to the reservoirs' volume (Junta de Vigilancia del Río Choapa y sus Afluentes (JVRC), 2015).

The contribution of the WF to water resources management: discussion of the constraints in the Coquimbo Region

The Coquimbo Region has focused its agricultural development on fruit crops, mainly table grapes, walnut and citrus. These crops are adapted to the prevalent soil conditions, temperatures and atmospheric humidity of the region; however, they can only be grown with irrigation because of the growing period, which is in spring and summer, when the precipitation is zero. The most important fruit crops in Limarí Basin are presented in Figure 3.

Even though the basin reservoirs provide irrigation security, water storage in the basins is not always enough to satisfy the total demand which can be observed during drought periods, for example, the 1995/96 season (Figure 3). These conditions where

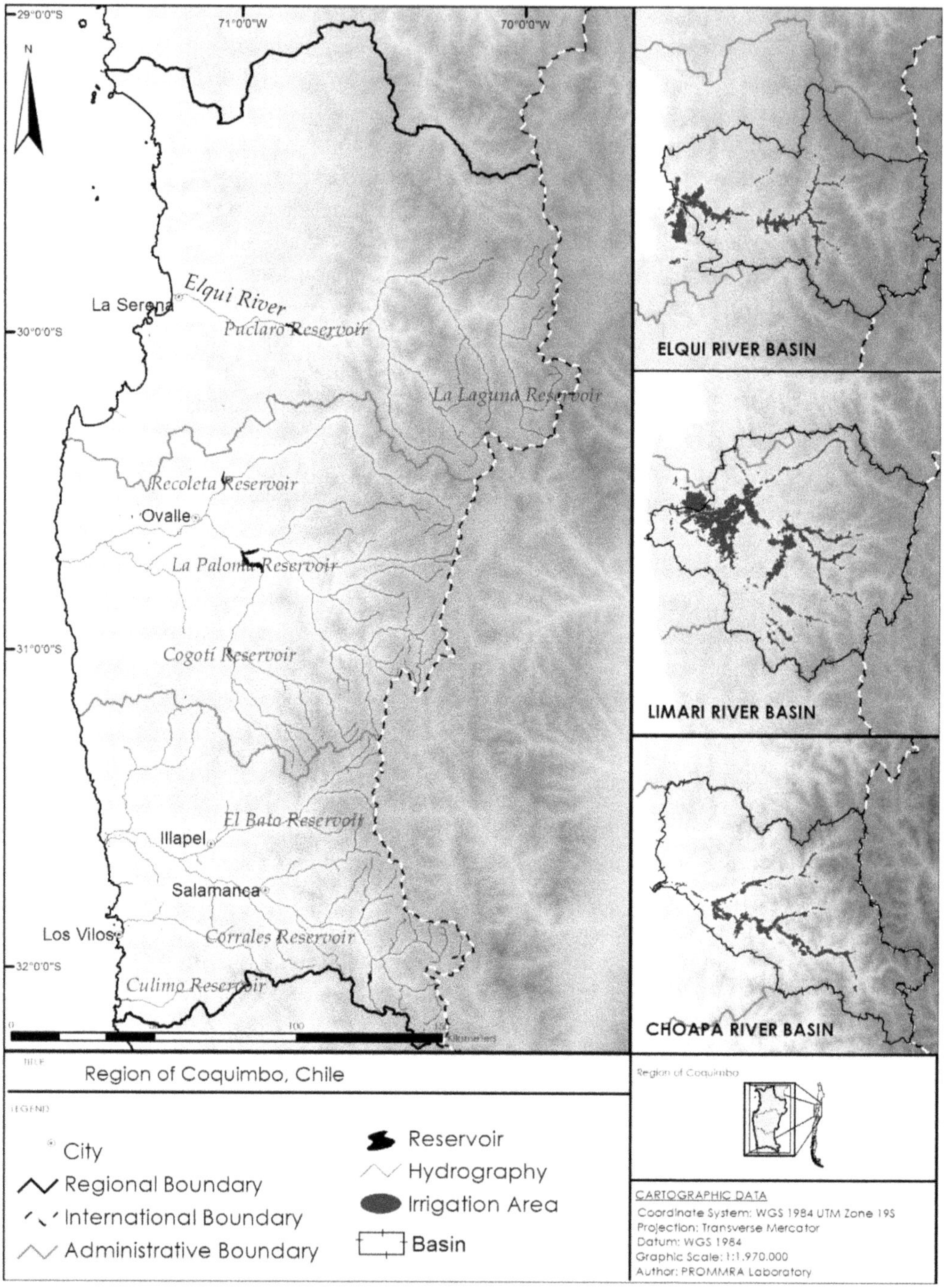

Figure 2. The Coquimbo Region and the Elqui, Limarí and Choapa basins.

the water distribution works according to water availability and private organization decisions force irrigation management to be more efficient. Vegetables irrigated with traditional systems show a seasonal variation of area during the period of analysis and a reduction during the 1995/96 season. On the other hand, the group of fruit trees is

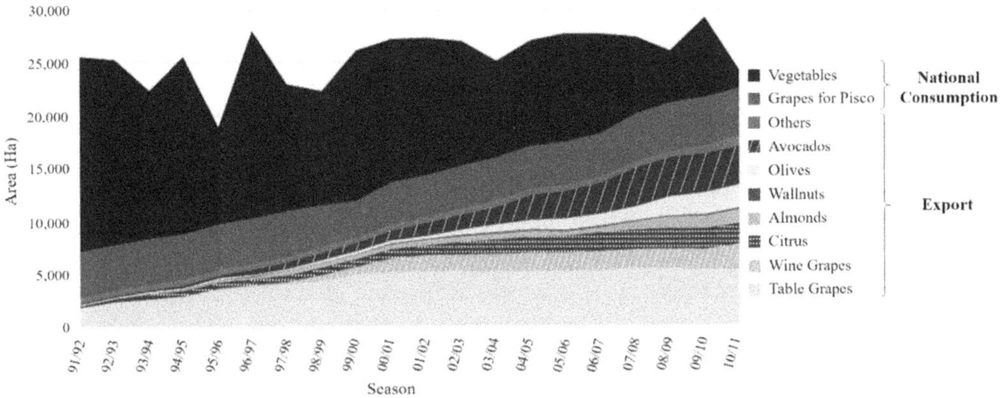

Figure 3. Evolution of the cultivated area according to the kind of crop in Limarí Basin for the 1991/92–2010/11 seasons.

irrigated by drip or micro-sprinkler irrigation and shows a permanent increase of area (Table 1). The water allocation regime as a flow per water right (l/s per water right) or as a water volume per water right for each season (m^3 per water right per season) has an effect in the relation between group of crops and irrigation technology.

The water consumption per hectare cannot be regulated by private or governmental organizations. In consequence, irrigation security depends on the balance between the water allocated per water right and the water demand per water right.

In Chile, the ownership of water rights does not depend on the ownership of the land; therefore, the amount of water rights per hectare is not a standard number. In consequence, greater irrigation security can be reached by buying water rights in the market or because of water demand reduction.

Demand reduction can be reached by technological improvements or diminishing the irrigated area. Technological improvements require a higher investment, while the surface reduction diminishes income, which is why the reduction of the hydro-climatic and financial risks is related to water allocated per water right, price of water rights, technological improvement and the agricultural product market.

The private property of water rights, and the water distribution related to the availability, show that the BWF could be an index that promotes by zone and crop the efficient use of resources reducing water demand or increasing the yield. However,

Table 1. Percentage of the area cultivated using different technologies during seasons 1995/96 and 2005/06.

	Season 1995/96		Season 2005/06	
Irrigation methods	Fruit trees (%)	Vegetables and annual crops (%)	Fruit trees (%)	Vegetables and annual crops (%)
Traditional	39.7	60.3	18.7	81.3
Micro-irrigation plus sprinkler	98.4	1.6	98.0	2.0

Sources: Results were obtained using a database from the censuses for 1997 and 2007 (INE, 1997, 2007).

in the case of Chile, the water rights market, and the private property regime itself, influences the rational use of water. The question is: are the market and the economic value of water rights enough to promote sustainable usage?

The WF corresponds to the water consumed in the whole agricultural production process, which is aggregated and divided into product kilograms (Parada-Puig, 2012). Considering this concept, it is possible to use a 'demand' index that would contribute to establishing the investment required (water rights) for an agricultural project.

Some methods used for calculating the WF consider the inflow, outflow and storage changes in the irrigation zones or basins. There are some methods that show the impact on water resources when referring to a production-zone characterization factor and to the existing crops in an area, whereas different methods are only based on the consumptive use of water. Regardless of the differences in quantification, the outcome of WF determination process must be significant for the water resources managing organizations, state organizations and farmers (Herath et al., 2013).

The methodologies, approaches and indicators to evaluate the impact on water usage are continuously changing. The WF as a concept has been a step; however, the WF calculation does not encompass water availability or shortages, which are often a concern in the impact evaluation of water consumption at basin level. In order to obtain WF data under the analysis zone's conditions, the water consumption impacts must be considered. They are best represented if the shortage factors from the arid zones are considered (Jeswani & Azapagic, 2011).

In the Coquimbo Region the availability of water resources for irrigation is limited, and therefore it is necessary to characterize crop water demand. The WF establishes the water needed per kilogram of product. The Water Code and the Chilean political constitution establish freedom of use for private water rights. In this context, the WF concept can be useful as a suggestion subordinate to private criteria.

While in arid zones at the basin scale the 'regular availability condition' has a low probability of occurrence, on an individual farmer scale in Chile it is possible to reach a higher stability of water availability with a higher number of water rights per hectare. The hydro-climatic variability in the arid zones of Chile and the water distribution per water right criterion both promote efficient water usage. However, currently, crop choices depend on different criteria and conditions such as capital to invest, climatic conditions and prices of agriculture products.

The WF can be an indicator that guides crop choice and associated technologies; however, in the arid context the efficiency/volume relationship applied to the irrigation process is variable. According to this idea, it is possible to identify the need to develop the WF as a concept in which hydro-climatic variability is included.

The context of the water market and private property of the water rights includes the economic investment in 'water rights' for agricultural projects, and thus the water consumption and crop yield are related to crop choice.

It is possible to observe in Figure 3 that most of the crops in Limarí Basin are produced for export. It also shows the increase of the cultivated area with fruits, and in consequence, increased dependence on water availability. The water usage competition in agriculture has guided the arid basins towards an 'agricultural export' system, with permanent water demand and reduced resilience (Vicuna, Alvarez, Melo, Dale, & Meza, 2014).

Table 2. Water footprint for the main crops in the Coquimbo Region (l/kg).

		Elqui River Basin			Limarí River Basin			Choapa River Basin		
		Upper river basin	Middle river basin	Lower river basin	Upper river basin	Middle river basin	Lower river basin	Upper river basin	Middle river basin	Lower river basin
Table grapes	Thompson Seedless	547.4	524.4	399.1	547.4	521.1	426.9	547.4	470.3	547.4
	Red Globe	467.4	463.3	342.7	467.4	445.3	357.7	467.4	402.6	467.5
Pisco grapes	Moscatel Rosada	396.9	453.9	358.4	578.3	536.4	329.3	467.4	365.0	377.7
	Pedro Jiménez	433.0	331.1	233.0	412.7	387.8	225.2	343.7	266.9	343.7
	Moscatel de Austria	284.2	317.7	257.7	407.6	409.3	284.0	391.2	321.5	348.1
Olives	Sevillano	872.2	901.4	761.9	–	–	–	–	–	–
	Arbequina	456.1	470.7	401.1	–	–	–	–	–	–
Avocado (tree)		462.6	477.7	436.9	462.6	477.7	358.1	462.6	477.7	462.6
Mandarin (tree)		251.1	258.8	220.8	251.1	263.4	195.6	251.1	222.5	251.1

Source: Osorio-Ulloa (2013).

Research from the Farming Investigation Institute (Instituto de Investigaciones Agropecuarias) in the Coquimbo Region includes estimations of the BWF reference values for the main crops (Table 2).

The reference values allow the relation between the irrigation water usage and the efficiency in the region. In the case of the table grape vine, the region shows a BWF of about 400 l/kg of fruit; pisco grape vine, 327 l/kg; avocados, l/kg; 378.6; olives, 537.7 l/kg; and citrus, 143.2 l/kg (Osorio-Ulloa, 2013).

A study on the main agricultural crop elaborated by the Regional Corporation of Productive Development of the Coquimbo Region shows that, for example, in the Red Globe variety of table grapes, the volumes applied fluctuate between 4000 and 9000 m^3/ha/season, leading to a production between 19,435 and 29,575 kg of fruit/ha/season. Analyzing the applied water productivity, table grapes obtain a water productivity between 350 and 144 l/kg. A similar situation was described for pisco grapes, avocado, mandarin and olive trees.

The spatial distribution of table grapes, pisco grapes, avocado, mandarin and olive trees in the basin is concentrated in downstream reservoirs with similar climatic conditions. This means that large variations in water productivity for the same crop could be related to irrigation technology and crop management.

The BWF reference values distributed in order of crops' location (high, medium and low areas) show a variability in the basins of the Coquimbo Region (Elqui, Limarí and Choapa) that is lower than variations in water applied and water productivity.

The effect of legislation and markets

The Chilean Water Code classifies water rights according to the water source, treating surface and ground waters as different. The legal condition affects the access to water rights, strengthening private property and the market. There are two markets related to water: the water rights market and the spot market of water volumes. In the first, water rights can be traded and in consequence there is a change of ownership. In the second, the price is related to the short-term (seasonal) water distributed per surface water right and to the seasonal irrigation demand. This market allows the adaptation to seasonal hydro-climatic variation. Groundwater rights are private and act as a secondary source in order to compensate for a surface water deficit during drought periods, because of the energy price associated with pumping groundwater.

The increase of conveyance and irrigation efficiency generates an increase in water availability per water right, which produces three private options for individual decisions: (1) increase cultivated area; (2) sell a water volume in the spot market; and (3) store water in big private reservoirs (i.e., La Paloma, Recoleta, Cogoti) using private accounts for future seasons.

Current public policies subsidize irrigation technologies and channel covering, which improve efficiency and generate an increase in water availability; consequently, they enlarge the irrigated area but generate a bigger area of perennial crops (fruit trees) with a long-term water demand, reducing the resilience during dry years.

In the case of Limarí Basin, the allocation rule is calculated for an 85% probability, meaning that in 15% of annual cases the allocation of water is not enough to complete a normal volume per water right per season. During dry years, the water allocated is lower than 40% of the normal volume for season (Ingenieros Civiles, 1978). As a consequence, depending on the water rights per hectare, perennial crops can have a 15% of probability of shortage throughout their entire productive life.

The second scenario is related to the spot market, in which the water volume allocated per water right can be appraised and traded under shortage conditions. It can help to reduce the effect of a shortage period in the yield of perennial crops, but the consequence is a reduction in vegetable cultivation (Figure 3). The water market is private and unregulated. The short-term water demand in the spot market enables the acquisition of water volume, as opposed to water rights.

The last scenario defines the alternative of storing a volume of water, which is the result of irrigation efficiency increase of the crop. This alternative is understood not as the water storage inside the farm or in the aquifers, but in the 'individual account' of each water right owner.

The water volume to allocate for each organization is communicated at the beginning of each season. Simultaneously, each organization generates 'individual accounts' for its shareholders.

In the context of the Water Code of 1981, the ratio of water rights to irrigated hectares is not a constant value. That is the reason why the water availability to irrigate a crop in a farm will depend on the number of water rights of the owner and the equivalent expected (85% probability) volume of water available per season. The perennial crops in the region need water irrigation every year. Therefore, the estimation of water rights per hectare for these crops is a challenge because it depends partly on the age of the plants, and there is a strong impetus to water at a minimum level needed to keep established perennial plants alive. An underestimation will increase the risk of shortage and an overestimation will lead to more investment in irrigation projects.

In Limarí Basin, the irrigated area of fruit trees has grown even during drought years, which means an increase in the permanent use of blue water resources to avoid a yield reduction or replanting of dead orchards.

The Central Valley in California (USA) has been described as having a similar evolution concerning land use (fruit trees) and blue water demand; nevertheless, in California the agriculture relies mainly on local groundwater, while the Coquimbo Region relies on surface water. In the same valley, during recent drought periods (2007–09 and 2012), groundwater pumping increased owing to a combination of decreased surface water availability and land-use changes (Faunt, Sneed, Traum, & Brandt, 2016).

Basin sustainability using the water footprint?

Water sustainability is a concept that could be related to BWF by comparing it with the blue water availability in the basin. If the WF is greater than the annual resource replenishment, then a blue water shortage is produced in the basin.

The application of this index depends on the ability to determine the real supplies and demands for each basin, which can be affected by information availability about hydrology (total runoff or water availability) and the private management of the basin.

The operational rules in the basins of the Coquimbo Region allow the distribution in the long term of a volume or flow per water right taking in account a probability of 85%. It means there is information about water availability when planning in the long term for the choice of crops. In Limarí Basin, between 1990 and 2010, the area cultivated with water-intensive crops for exportation increased, whereas the area of vegetables was reduced. The water allocation system based in operational rules does not promote a balance between different water-intensive crops to improve sustainability.

The water availability, land use and crop productivity are not uniform in the basin and it is necessary to obtain the historical water balances and WF to develop the local water sustainability indices (Chapagain & Orr, 2008).

The variability of WF data in the Chilean context can be affected by geographical and climatic differences; differences between varieties of the same crop; and the frequency of shortage periods. Yield information is private, therefore the access to this data is limited. In consequence, field research and the use of remote sensing are necessary to improve the knowledge of local BWF (Yang & Zehnder, 2007).

The BWF shows the water volume used in the basin, but it is also conditional on water-use efficiency at the field level. In the Coquimbo Region, 49.2% of the irrigated area uses pressurized irrigation systems, such as drip irrigation, micro-spraying and traditional spraying (Instituto Nacional de Estadística (INE), 2007). In the Coquimbo Region, almost the total area of fruit trees is irrigated with pressurized systems and the water availability in the long term is determined by the rules of operation; therefore, the increase of water-intensive crops in the area will reduce the area of vegetables or annual crops. In consequence, the basin's sustainability, especially during drought periods, will be reduced.

In the Coquimbo Region the implementation of pressurized irrigation systems and monitoring technology to calculate the actual ET have been improved, but the development of an integrated system in order to calculate the blue water demand and crop water consumption is a challenge. On the other hand, there is local information about the BWF; therefore, it could be possible to integrate the blue water balance and the BWF to develop a local index of sustainability.

The actual ET and the potential yield of crops related to the BWF allow an estimate to be made of the local gaps between the real situation and an expected situation. The reduction of these gaps could guide public policies promoting technologies to reduce the ET (nettings over the canopy of fruit trees) and not only to subsidize the irrigation technology. Also, the public policies can promote technologies to improve the yield of water-intensive crops to reduce the BWF. Particularly in fruit trees, the reduction of time between planting season and the beginning of harvest can be stimulated.

In the Chilean arid zone, the BWF of crops and the irrigation demand coverage is determined mainly by the water supply, the amount of water rights and the agronomic management. A large increase in crop efficiency, without a rise or reduction in ET, can be achieved through the proper management of nutrients, water and soil, providing higher water productivity (Mekonnen & Hoekstra, 2014).

In Limarí Basin, the total irrigated area varies year by year, but does not show an upward trend. However, the area of water-intensive crops increases permanently. As a consequence, the principal problem is basin sustainability according to crop composition. The competition for water rights and volumes between water-intensive and less intensive crops without including an index of sustainability reduces the resilience in the basin.

The private organizations (OUAs) need local information to promote the sustainability in their zone of influence, including water security and BWF information. They can allocate the water by using the operational rules, but also can alert about crop composition by taking into account the availability of blue water and the risk of drought.

Knowledge about the relation of water rights per hectare to ensure crop yield in the long term is necessary to guide decisions, such crop choice and the amount of water rights to invest. Information about the BWF in the Chilean arid zones has to be included to promote sustainable agriculture development.

In the Chilean arid zones, the water supply actually limits the increase in total demand and the market guides the water-right owners' decisions; nevertheless, the information for planning a sustainable private investment could be promoted by the OUAs.

Conclusions

The arid condition of the Coquimbo Region limits water availability. The OUAs follow operational rules and private criteria to allocate volumes or flows per water right by considering long term a probability of 85%.

The total cultivated area does not show an increase, but it varies from year by year. Furthermore, the area cultivated with fruit trees, including water-intensive crops, shows a permanent upward trend. The consequence of this is the reduction of vegetables area. The water allocated in the long term limits the total irrigated, but it is not enough to promote a sustainable land use. The fruit trees for exportation are irrigated with pressurized systems and require a permanent water supply. Therefore, during drought periods, when water allocation is reduced (15% of probability), productivity and survival will be severely affected.

The BWF and the monitoring of blue water availability could be useful indicators to improve the sustainability of water resource management in the basins of the Chilean arid zones. These indicators require local and historical information of water allocation decision-making (m^3 per water right per season) located in water right owners' organizations (OUAs). Additionally, the Coquimbo Region has some data about WF and BWF that have to be improved. As a consequence, is necessary to improve the local research to estimate long-term blue water availability and improve the knowledge about and monitoring of BWF. These indicators can be useful for the sustainability of investment.

The public policies promote principally the private investment in irrigation efficiency, but not in the stimulation of planning to reduce the WF. A challenge of water management in the Coquimbo Region is to integrate sustainability criteria into public policies and private water management, including farmers and OUAs.

It is necessary to improve public and private policies to develop sustainable indicators related to blue water balance. These indicators require a monitoring system for both land use and the real yield of crops in the basin.

The organizations are not authorized to modify how the water rights are used; nevertheless, they can transfer information to farmers about the long-term blue water balance to improve water security. The challenge for the OUAs is to improve their capacities.

Efficiency and technology determine the relationship between the WF and system sustainability. In the arid zone of Chile, an efficiency increase provides a higher net water availability for the water rights' owner. This promotes a rise in water demand and speculation about water volumes in the spot market. It is necessary to develop distributed or semi-distributed models to simulate different scenarios, including changes in land use to determine long-term conditions to maintain a positive or neutral blue water balance.

Acknowledgments

This paper was given at the workshop on Virtual Water in Agricultural Products: Quantification, Limitations and Trade Policy (Lincoln, Nebraska, USA, 14-16 September 2016), which was sponsored by the OECD Co-operative Research Programme: Biological Resource Management for Sustainable Agricultural Systems whose financial support made it possible for the author to participate in the workshop.

Disclosure statement

No potential conflict of interest was reported by the author.

Funding

This work was supported by the Co-operative Research Programme: Biological Resource Management for Sustainable Agricultural Systems, Organisation for Economic Co-operation and Development (OECD) [grant number TAD/CRP JA 87788].

References

Allan, J. A. (1998). Virtual water: A strategic resource. Ground Water, 36(4), 545–547.
Bauer, C. J. (1997). Bringing water markets down to earth: The political economy of water rights in Chile, 1976–95. World Development, 25, 639–656.
Chapagain, A. K., & Orr, S. (2008). An improved water footprint methodology linking global consumption to local water resources: A case of Spanish tomatoes. Journal of Environmental Management, 90, 1219–1228.
Comisión Nacional de Riego (CNR). (2004). Diagnóstico base de los embalses El Bato y Corrales, IV Región. Centro de Información de Recursos Naturales: Retrieved from http://bibliotecadigital.ciren.cl/bitstream/handle/123456789/9603/CNR-0126_3.pdf
Corporación Nacional del Medio Ambiente (CONAMA). (2002). De Mar a Cordillera – Región de Coquimbo [From sea to mountain range – Coquimbo Region]. Sistema Nacional de Información Ambiental. Retrieved from http://www.sinia.cl/1292/articles-29099_recurso_2.pdf
Díaz, R. R. (2008). El recurso hídrico: Acceso, reglas y restricciones en zonas áridas, Cuenca del Limarí, Chile(Unpublished bachelor's thesis). Universidad de La Serena, La Serena, Chile.
Dirección General de Aguas (DGA). (2004a). Diagnóstico y Clasificación de los cursos y cuerpos de agua según objetivos de calidad – Cuenca del Río Elqui. Sistema Nacional de Información Ambiental. Retrieved from http://www.sinia.cl/1292/articles-31018_Elqui.pdf
Dirección General de Aguas (DGA). (2004b). Diagnóstico y Clasificación de los cursos y cuerpos de agua según objetivos de calidad – Cuenca del Río Limarí. Sistema Nacional de Información Ambiental. Retrieved from http://www.sinia.cl/1292/articles-31018_Limari.pdf
Dirección General de Aguas (DGA). (2004c). Diagnóstico y Clasificación de los cursos y cuerpos de agua según objetivos de calidad – Cuenca del Río Choapa. Sistema Nacional de Información Ambiental. Retrieved from http://www.sinia.cl/1292/articles-31018_Choapa.pdf
Faunt, C. C., Sneed, M., Traum, J., & Brandt, J. (2016). Water availability and land subsidence in the Central Valley, California, USA. Hydrogeology Journal, 24, 675.
Herath, I., Green, S., Singh, R., Horne, D., van der Zijpp, S., & Clothier, B. (2013). Water footprinting of agricultural products: A hydrological assessment for the water footprint of New Zealand's wines. Journal of Cleaner Production, 41, 232–243.
Ingenieros Civiles, B. F. (1978). Operación del Sistema Paloma para Ministerio de Obras Públicas, Dirección General de Obras Públicas. Dirección de Riego.
Instituto Nacional de Estadística (INE). (1997). Censo Agropecuario y Forestal [on line]. Retrieved from http://www.censoagropecuario.cl/
Instituto Nacional de Estadística (INE). (2007). Censo Agropecuario y Forestal [on line]. Retrieved from http://www.censoagropecuario.cl
Jeswani, H. K., & Azapagic, A. (2011). Water footprint: Methodologies and a case study for assessing the impacts of water use. Journal of Cleaner Production, 19, 1288–1299.
Junta de Vigilancia del Río Choapa y sus Afluentes (JVRC). (2015). Memoria Hídrica Junta de Vigilancia del río Choapa y sus Afluentes. JVRC. Retrieved from http://www.jvriochoapa.cl/wp-content/uploads/2014/11/MEMORIA-2015-JVRCH.pdf
Mekonnen, M. M., & Hoekstra, A. Y. (2011). The green, blue and grey water footprint of crops and derived crop products. Hydrology and Earth System Sciences, 15, 1577–1600.
Mekonnen, M. M., & Hoekstra, A. Y. (2014). Water footprint benchmarks for crop production: A first global assessment. Ecological Indicators, 46, 214–223.
Osorio-Ulloa, A. (2013). Determinación de la huella del agua y estrategias de manejo de recursos hídricos [Determination of water footprint and water resources management strategies]. La Serena, Chile: Instituto de Investigaciones Agropecuarias, Centro Regional de Investigación Intihuasi.
Parada-Puig, G. (2012). El Agua virtual: conceptos e implicaciones [Virtual water: Concepts and implications]. Orinoquia, 16, 69–76. Retrieved from http://www.scielo.org
Pellicer, F., Perni, A., & Martinez, J. M. (2013, December). La Huella Hídrica total de una Cuenca: El caso de la Demarcación Hidrográfica del Segura. Paper presented at the meeting of VIII Congreso Ibérico sobre Gestión y Planificación del Agua, Lisbon.

Qin, L., Jin, Y., Duan, P., & He, H. (2016). Field-based experimental water footprint study of sunflower growth in a semi-arid region of China. Journal of the Science of Food and Agriculture, 96, 3266–3273.

Universidad de Los Lagos. (2009). Experiencias Nacionales e Internacionales de Articulación de Actores. Estudio Articulación y Actores para la Descentralización en Tres Regiones de Chile. Programa Ciudadanía y Gestión Pública. Retrieved from Subsecretaria de Desarrollo Regional y Administrativo: http://www.subdere.gov.cl/sites/default/noticiasold/articles-82969_recurso_29.pdf

Vicuna, S., Alvarez, P., Melo, O., Dale, L., & Meza, F. (2014). Irrigation infrastructure development in the Limarí Basin in Central Chile: Implications for adaptation to climate variability and climate change. Water International, 39(5), 620–634.

Vivanco, C. (2014). Operatividad del sistema Paloma en base a criterios de asignación hídrica, aplicado al modelo WEAP-Limarí (Unpublished bachelor's thesis). Universidad de La Serena, La Serena, Chile.

Yang, H., Pfister, S., & Bhaduri, A. (2013). Accounting for a scarce resource: Virtual water and water footprint in the global water system. Current Opinion in Environmental Sustainability, 5, 599–606.

Yang, H., & Zehnder, A. (2007). 'Virtual water': An unfolding concept in integrated water resources management. Water Resources Research, 43, W12301.

Zavala, H., & Trigos, H. (2008). Hidrología de la Cuenca del Río Elqui [Hydrology of the Elqui River Basin]. In P. J. Cepeda (Ed.), Los Sistemas Naturales de la Cuenca del Río Elqui (Región de Coquimbo, Chile): Vulnerabilidad y cambio del clima [The Natural Systems of the Elqui River Basin (Coquimbo Region, Chile): Vulnerability and Climate Change] (pp. 66–164). La Serena: Ediciones Universidad de La Serena.

The effect of diet changes and food loss reduction in reducing the water footprint of an average American

Mesfin M. Mekonnen and Julian Fulton

ABSTRACT

We analyzed the consumptive water reduction of diet shifting along with food waste reductions in the case of the United States. We find that a dietary shift to healthy diet will not always lead to reduced water footprint. Dietary shifts to vegan and vegetarian diets provide larger reduction in the consumptive water footprint. Reducing food loss and waste produced the largest potential water footprint reduction in our analysis of the US food system. Our findings suggest that a combination of measures that include dietary shift, reducing caloric intake, and reducing food waste result in a significant decline in water footprint.

Introduction

A growing world population, which is expected to reach 9 billion by 2050, combined with an increase in per capita income and per capita caloric intake will increase the global food demand, particularly the demand for meat, dairy and processed food products (Godfray et al., 2010; Tilman & Clark, 2014). These dietary behaviors are associated with health problems such as overweight and obesity among the population, thereby increasing the healthcare cost of many countries. Environmental costs are similarly increasing, as evidenced by the competition for scarce fresh water, energy, land, and the increase in greenhouse gas emissions. Therefore, providing a healthier diet to a future population of 9 billion with minimal environmental impacts is a major challenge facing the global food production systems. This challenge can be addressed through a combination of methods that include sustainable intensification to close the yield gap (Foley et al., 2011; Tilman, Balzer, Hill, & Befort, 2011), change in the dietary behavior (Foley et al., 2011; West et al., 2014), and reducing food waste (Kummu et al., 2012; Vanham, Bouraoui, Leip, & Bidoglio, 2015).

Recent research has shown the effect of dietary shifts on greenhouse gas emissions (Meier & Christen, 2013; Tilman & Clark, 2014), on energy use (Meier & Christen, 2013), on land use (Tilman & Clark, 2014), and on water footprint (Jalava, Kummu, Porkka, Siebert, & Varis, 2014; Vanham, Mekonnen, & Hoekstra, 2013a). What is

---

The opinions expressed and arguments employed in this publication are the sole responsibility of the authors and do not necessarily reflect those of the OECD or of the governments of its Member countries.

Supplemental data for this article can be accessed at https://doi.org/10.1080/02508060.2018.1515571.

common in all of these studies is that a shift to healthier, vegetarian diets put the smallest pressure on the scarce water resources.

A number of countries including the US are putting forth a concerted effort to raise citizen awareness to adopt healthy diets. One such effort is the *Dietary Guidelines for Americans* (HHS & USDA, 2015), which is published by the US Department of Health and Human Services (HHS) and the US Department of Agriculture (USDA). The guidelines provide healthy-diet recommendations for different age groups and activity levels. It further provides Mediterranean, vegetarian, and vegan dietary recommendations. These guidelines, however, do not address the potential reductions in water and other resources use associated with diet-shifting, which is the central concern of our study.

The main objective of our study is to assess the consumptive (green and blue) water footprint of the different *Dietary Guidelines for Americans*-recommended diets. The water footprint is an indicator of the direct and indirect water use to produce the goods and services we use (Hoekstra, Chapagain, Aldaya, & Mekonnen, 2011). Only agricultural food products are considered in this study. The blue water footprint is the water from surface and groundwater that is evapotranspired by agricultural crops or incorporated into the product during growth and processing. The green water footprint is the rainwater that is stored in the root zone of the soil and evapotranspired or incorporated into the product.

Materials and methods

Diets configuration

In this study we evaluated six different diets – the USDA Loss Adjusted Food Availability (LAFA), USDA recommended healthy diet (HEALTHY), healthy diet at 2,000 kcal/day (HEALTHY2k), Mediterranean diet (MEDIT), vegetarian diet (VEGY), and a VEGAN diet for an average American. The LAFA diet represents the current average American diet and is used as the 'least healthy' alternative. It was obtained from USDA-ERS (2017). The HEALTHY, HEALTHY2k, MEDIT and VEGY diets configurations were obtained from the 2015–2020 *Dietary Guidelines for Americans* by HHS & USDA (2015), and the VEGAN diet was taken from *Dietary Guidelines for Americans* 2010 by USDA & HHS (2010). The 2015 LAFA database was missing data on the 'added fats and oil' group so we adopted the 2010 food intake value for this group to fill the missing gap. The total kcal intake for the LAFA diet for 2015 was 2423 kcal. The *Dietary Guidelines for Americans*, on the other hand, provides different food intake patterns from 1000 to 3200 kcal for different age groups (HHS & USDA, 2015). Therefore, we selected the daily food intake pattern for 2400 kcal for the HEALTHY, MEDIT, and VEGY diets to be comparable to the LAFA food intake level. The HEALTHY2k diet was adopted from the Dietary Guidelines to assess the effect of dietary shift as well as the reduced caloric intake (at 2000 kcal/day instead of the 2400 kcal/day) on the consumptive water footprint. The level of detail in the two databases differ: while the per capita Calorie consumption and food pattern for the LAFA are provided for over 200 agricultural commodities, the *Dietary Guidelines for Americans* provide broadly aggregated food groups and sub-groups.

Food availability and loss in the LAFA database

The LAFA diet was constructed from the food availability data after accounting for the food loss through spoilage, plate waste, and other losses to represent the actual consumption of an average US consumer. According to Buzby, Hodan, and Jeffrey (2014) in 2010 in the United States, about 31% or 1249 kcal/cap/day of the food produced with total estimated value of $161.6 billon was wasted, spoiled, or otherwise not consumed by humans. Hence the estimated food loss in LAFA could actually be much greater than deduced here.

Table 1 presents the mass and caloric content of food groups actually consumed and the food loss at retail and consumer levels. Out of the 3660 kcal/cap/day food that was available for human consumption, only about 66% was actually consumed, while the rest was lost at the retail level (12%) and consumer level (22%). The largest Calorie loss was due to fats and oils (29%), followed by grains (18%), and sugar & sweeteners (20%). In terms of grams of food loss (g/cap/day), the largest loss was due to vegetables (23%), followed by fruits (19%) and dairy products (16%).

The water footprint of different diets

To estimate the water footprint (WF) of the different diets, we followed the approach defined by the Global Water Footprint Standard and developed by the Water Footprint Network (Hoekstra et al., 2011). The WF of diet d (WF_d, liters per capita per day) was calculated as:

$$WF_d = \sum_p (C_d[p] \times WF^*[p]) \tag{1}$$

where $C_d[p]$ is the volume of food product p intake under diet d (in kg/cap/day), the $WF^*[p]$ is the national average water footprint of food product p (in L/kg). The water footprint of the different food items was obtained mainly from Mekonnen and Hoekstra

Table 1. Foods actually consumed and food loss at retail and consumer level based on the LAFA database for 2015.

		Lost food			Lost food	
	Consumed	Retail level	Consumer level	Consumed	Retail level	Consumer level
Food group	(g/cap/day)			(kcal/cap/day)		
Vegetables	194	28	107	131	14	49
Fruits	140	24	89	82	13	54
Dairy products	204	33	60	219	27	70
Red meat	89	5	28	236	14	74
Poultry	73	4	16	166	8	36
Grains	151	26	39	524	90	135
Eggs	24	4	13	38	6	20
Fish	12	2	6	13	2	7
Nuts	13	1	1	75	5	8
Fats and oils	71	23	19	580	197	158
Sugar & sweeteners	95	17	48	360	67	182
Total	1066	166	427	2423	444	793

Data source: USDA-ERS (2017)

(2011a), who provided a weighted average WF for a large number of products for different countries including the US. This weighted average WF of the different agricultural products was calculated with the assumption that the item consumed originates from domestic production and imports according to their relative volumes. When data on the WF of the food item is missing from Mekonnen and Hoekstra (2011a), we used national average values from Mekonnen and Hoekstra (2011b) for crop products and from Mekonnen and Hoekstra (2012) for the animal products.

The volume of consumption under the five diets (HEALTHY, HEALTHY2k, MEDIT, VEGY, VEGAN) from the *Dietary Guidelines for Americans* are provided on aggregated food groups or sub-groups. Therefore, we used the detailed food items provided for the LAFA diet to derive an average WF for the different food groups (in L/kg) and used this to estimate the WF (L/cap/day) for the HEALTY, HEALTHY2k, MEDIT, VEGY, and VEGAN diets. The full list of the products consumed under the LAFA diet and the WF calculation is provided in the Supplementary Material.

Analysis of the different diets

The food intakes for the six diets are given in Table 2. The five recommended diets have close to two times more vegetables and fruits in their food composition compared to the LAFA diet. On the other hand, the LAFA diet is rich in vegetable oil, animal fats and sugar products compared to the other diets. Among the recommend diets, the HEALTHY diet and the MEDIT diet differ very little in product composition. The main differences between them are that in the MEDIT diet, the amount coming from other products, such as animal fat and sugar, is smaller, whereas seafood is a larger component compared to the HEALTHY diet. The quantity of vegetables, fruits, grains, and vegetable oils in the VEGY and VEGAN diets is the same or very close to that of the HEALTHY and MEDIT diets. The only difference is in the protein foods – VEGY diet replaces the meat and poultry products with more nuts, soybean and other legumes, whereas the VEGAN diet replaces all animal products with plant-based food items:

Table 2. Average daily food intake under the four diets (gram/cap/day)[a]

		HHS and USDA recommended American diet [b]				
Food Group	LAFA	HEALTHY	HEALTHY2k	MEDIT	VEGY	VEGAN[c]
Vegetables	194	379	323	379	379	379
Fruits	140	334	334	418	334	334
Grains	151	180	135	180	192	180
Dairy	204	405	405	338	405	405[d]
Protein Foods	211	190	160	214	125	183
Seafood	12	40	32	65		
Meats, poultry, eggs	186	133	112	133		
Nuts seeds, soy	13	16	16	16	52	81
Eggs					21	
Legumes (beans & peas)					51	102
Oils	45	31	27	31	31	21
Calories from Other Uses	115	77	59	66	85	72

[a]The units are converted to grams by using the conversion factor derived from LAFA data.
[b]Data from HHS & USDA (2015) 2015–2020 *Dietary Guidelines for Americans*.
[c]Data from USDA & HHS (2010) *Dietary Guidelines for Americans* 2010.
[d]Dairy for VEGAN diet is composed of calcium-fortified beverages and foods from plant sources.

animal-based dairy is replaced with plant-based dairy such as soy yogurt and beverage, rice milk, and tofu. The meat and egg products are replaced with more legumes. The recommended diets are specifically designed to reduce overweight and obesity and the associated health problems.

Results

The water footprint of the American diets

The water footprint of the different diets for the average American – LAFA, HEALTHY, HEALTHY2k, MEDIT, VEGY, and VEGAN – per food intake group are shown in Figure 1. The WF of an average American (LAFA) after adjusting for the food loss through spoilage, plate waste, and other losses was 1868 L per capita per day. A shift to recommended HEALTHY diet and MEDIT diet would increase the consumptive WF by 7% and 5%, respectively. Shifting to the VEGY and VEGAN diet would reduce the total consumptive WF by 20% and 37%, respectively. A HEALTHY diet at a lower level of caloric intake would also reduce the consumptive WF by 6%.

The WF reduction that was achieved by reducing the red meats, poultry and egg by shifting from the LAFA diet to the HEALTHY diet and MEDIT diets was overwhelmed by the increase in the dairy, fruits and vegetables. In terms of health benefits, the move from red meat to dairy, fruits and vegetables could be beneficial; however, the dietary shift does not help in reducing the consumptive WF. This finding is consistent with the findings by Tom, Fischbeck, and Hendrickson (2016) who found that the shift to a diet with a recommended food mix increased the average blue WF by as much as 16%.

The health and environmental benefits of dietary shift from the current diet (LAFA) to VEGY and VEGAN diet is clear from the analysis. Both diets provide a significant reduction in the consumptive WF. This is consistent with the findings from earlier studies (Vanham et al., 2016, 2013a).

Figure 2 presents the green and blue WF of the six diets. In all diets, the green WF accounts for the largest share, from 79% in the VEGAN to 90% in the LAFA diet. The effect of dietary shift on the relative green and blue WF differs among the different diets. Dietary shift from LAFA to the other diets increases the blue WF; shift to HEALTHY, MEDIT, VEGY, and VEGAN diet increases the blue WF by 35%, 40%, 31%, and 28%, respectively. While the shift to HEALTHY and MEDIT diet increases the green WF by 4% and 1%, respectively, shifting to HEALTHY2k, VEGY, and VEGAN diets reduces the green WF by 9%, 26%, and 44%, respectively. The major component of the increase in the blue WF comes from the increased consumption of fruits, grains and dairy products. The other reason for the increase in blue WF is the relatively larger blue WF per unit of product (L/kg) for fruits, which are mainly grown in the dryer western and southern part of the US. In these places, the available precipitation is not enough to meet the crop water requirements so irrigation (blue water) is applied to fill the deficit.

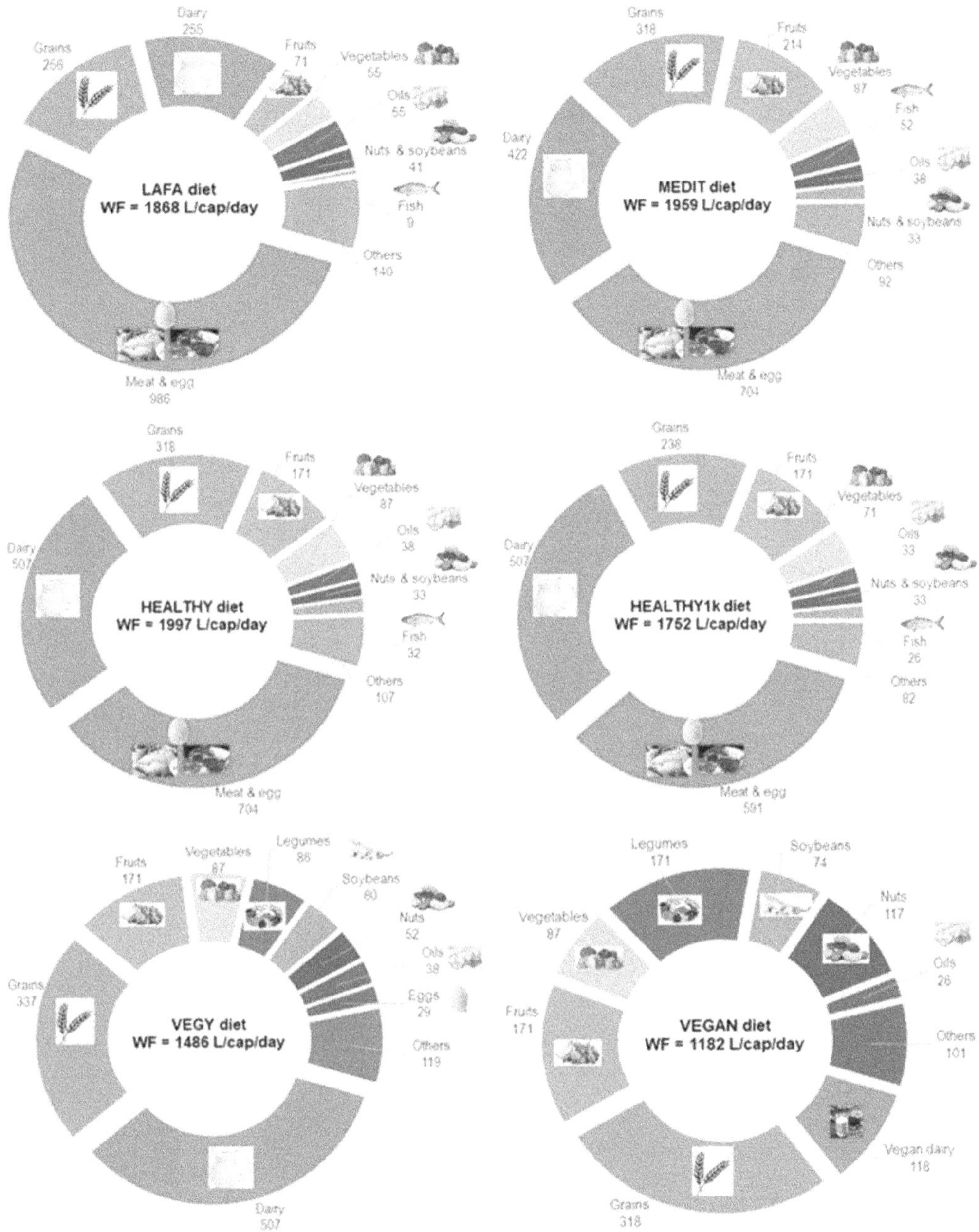

Figure 1. The consumptive water footprint (in L/cap/day) of the six diets for an average American.

Water footprint of food loss

We estimated the amount of consumptive water use that is associated with the food loss and waste in the US for the current LAFA diet. In 2015, about 34% or 1237 kcal/cap/day of the food available for consumption was lost or wasted (see Table 1 for the breakdown of the food loss). The breakdown of the WF of food loss in liters per capita

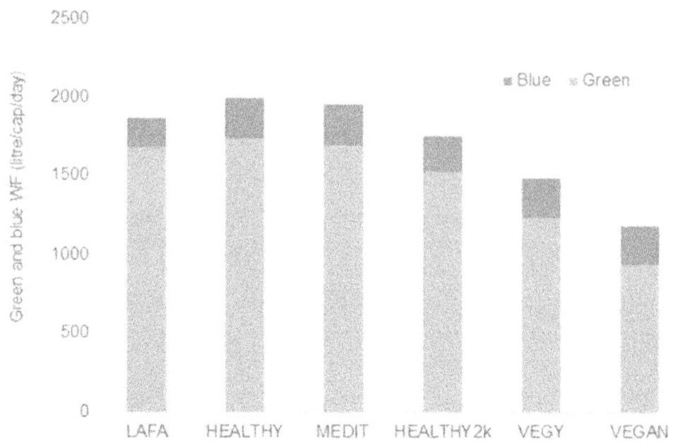

Figure 2. The green and blue water footprint of the six diets for an average American.

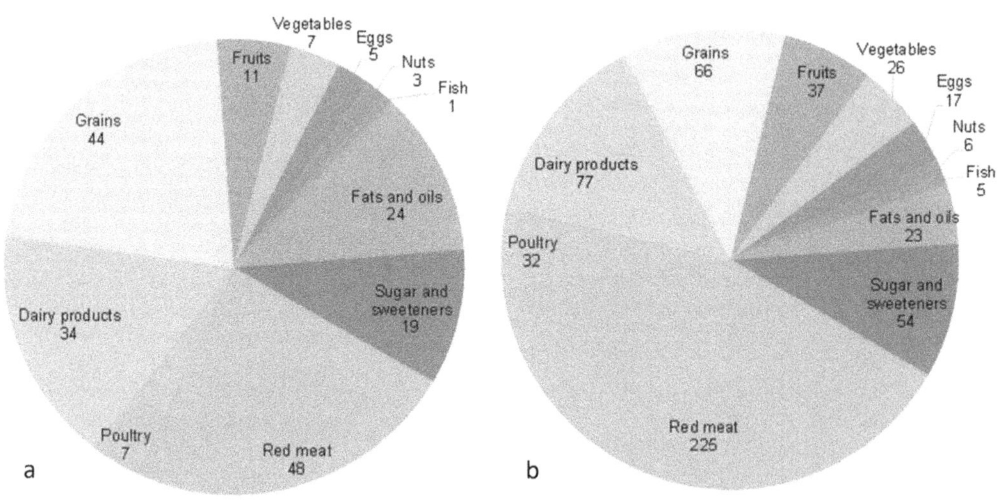

Figure 3. The consumptive WF of the wasted food (L/cap/day) per major product groups at retail level (a), and consumer level (b) from the LAFA diet.

per year is shown in Figure 3. The largest food loss (72%) and associated consumptive WF occurs at the consumer level. At both the retail and consumer levels, the largest WF related to food loss is due to red meat, dairy and grains. Vegetables and fruits account for the larger share in the volume of wasted food (g/cap/day), but because of its high water intensity (L/kg), red meat takes the largest share (53%) of the WF related to food loss.

In the US, a total of 90 km^3/y (88% green and 12% blue) of water was lost together with the food loss. This WF is slightly (12%) larger than the water saving that could be achieved by shifting from LAFA to the VEGAN diet (80 km^3/y), but twice as large as the potential water saving that can be achieved by shifting diet from the LAFA to VEGY diet. The latter shift could save about 52 km^3/y in green water, but cost about 7 km^3/y of more blue water consumption with a net saving (green plus blue) of 45 km^3/y.

Discussion

The dietary transition to more processed foods, refined sugars and fats, and meats has led to a widespread prevalence of overweight and obesity in many countries, foremost among US residents. These unhealthy eating habits increase diet-related diseases and national healthcare costs. According to HHS & USDA (2015), in 2008, obesity-linked medical costs were about $147 billion in the US. The *Dietary Guidelines for Americans* is designed to prevent the diet-related chronic diseases by promoting healthier diets among the population.

Our results show that while the HEALTHY diet could have health benefits, it can simultaneously increase the consumptive WF. This is because partial substitution of meat with increased intake of more-healthful food items such as fruits, vegetables, and legumes will increase the consumptive WF. This goes against the general expectation that there will be an automatic reduction in the WF if we substitute meat with other less water intensive food items. Other studies indicate that the MEDIT diet helps to reduce the WF (Blas, Garrido, & Willaarts, 2016). However, in our study both the blue and green WF increased by shifting from the LAFA to MEDIT diet. The major reason could be that the suggested MEDIT diet in the *Dietary Guidelines for Americans* (HHS & USDA, 2015) could be quite different from how other researchers compose the MEDIT diet. Shifting to VEGY diet reduces the green WF but increases the blue with an overall reduction of the consumptive WF. This was achieved by replacing meat fully by other, less water-intensive products. Further consumptive WF reduction is achieved by shifting to the VEGAN diet, which replaces all animal products with plant-based food items. The VEGY and VEGAN diets have both health and environmental benefits along with reduced consumptive WF, including reducing greenhouse gas emissions (Tilman & Clark, 2014).

The potential WF reduction by shifting to VEGY diet is quite small compared to what could be achieved by avoiding or reducing food waste. The total green and blue water that is lost through food waste is close to two times larger than the WF reduction that could be achieved by shifting to the VEGY diet and 12% larger than that is achievable by VEGAN diet. Considering that about 800 million people globally are undernourished (FAO et al., 2015), wasting huge amounts of food is unethical (Gjerris & Gaiani, 2013) as well as uneconomical and harmful to the environment. Therefore, dietary change should be combined with actions aimed at reducing food loss and waste in order to significantly diminish the water challenge facing nations and the world (Foley et al., 2011; Jalava et al., 2016).

Comparison with other studies

On a per capita basis, our estimate of consumptive WF is quite low compared to the values provided by Mekonnen and Hoekstra (2011a). The US per capita daily consumptive WF from Mekonnen and Hoekstra (2011a), after subtracting the items that are missing from the LAFA data (such as spices, coffee, tea, fiber products, tobacco, rubber, hides & skins), was 5118 L/cap/day. Their value is based on consumption i.e., caloric intake plus food loss and waste. Therefore, to compare the two estimates, we added the consumptive WF related to food loss and waste to the LAFA diet WF. The sum of the LAFA diet WF (1868 L/cap/day) and the WF related to food loss and waste was 2637 L/cap/day, which is still less than half the estimate by Mekonnen and Hoekstra (2011a). There are a number of reasons for the difference:

(i) The most important difference was that the WF of the added sweeteners used in the two study. In Mekonnen and Hoekstra (2011a), the green plus blue WF of the sweetners was 9848 l/kg while in the current study, the WF of the corn sweeteners was 801 l/kg. This large difference in the WF of the sweeteners has led to a lower per capita consumptive WF in the current study compared to the earlier one (by as much as 922 l/cap/d),
(ii) The per capita consumptive WF related to grains was smaller in the current study by 228 l/cap/d because of limited number of grains considered in LAFA data compared to the FAO's Food Balance Sheet (FBS),
(iii) There was also a difference in the per capita consumptive WF related to dairy and vegetable oils totaling 306 l/cap/d. The difference is mainly due to disparities in the weight of the products in the LAFA and FBS database,
(iv) Slightly larger meat consumption as reported in the FBS also contributed about 700 l/cap/d.

The food supply values reported by FBS for a number of products are consistently larger than those reported in HHS & USDA (2015) food availability at the retail level.

Our estimate of blue WF for the LAFA diet plus the blue WF for the food waste (33 km^3/y) is 18% smaller on average than that of Tom et al. (2016). This difference is due to variations in the food mix used in the two studies. While we have directly used the USDA data, Tom et al. (2016) have derived their own food mix which slightly differs from that provided in the LAFA database (HHS & USDA, 2015). There is good agreement between the two studies in the relative increase in the blue WF by shifting diet from LAFA to the USDA recommended HEALTHY diet – an increase of 13% in our estimate compared to 16% by Tom et al. (2016).

To sum up, while a number of studies have shown a significant reduction in the WF by shifting to healthier diets (Meier & Christen, 2013; Vanham, Hoekstra, & Bidoglio, 2013b) or the Mediterranean diet (Blas et al., 2016; Vanham et al., 2016), the current study found that shifting to HEALTHY diet or MEDIT diet would increase the WF by 7% and 5%, respectively. The difference in the final result can partly be explained by the fact that we used the loss adjusted food intake as a baseline, whereas the other studies used the full consumption, i.e., intake plus food loss. On the other hand, although the magnitude differs from study to study, all studies showed that dietary shift to VEGY and VEGAN diets will result significant WF reductions.

A significant amount of water is lost through food loss and wastage. The water savings through avoiding food loss and waste is as much as twice that could be achieved by shifting to VEGY diet. Therefore, a combination of dietary shift to VEGY and VEGAN plus reduction in food loss and waste at the consumer and retail levels would provide a healthier population and environmental benefits.

The current study focused only on water, but the effect of diets is multidimensional, requiring a holistic analysis that includes the impacts of other resources such as land

and energy, greenhouse gas emissions, nutrient loads to freshwater, and human health effects (Tilman & Clark, 2014; Tom et al., 2016).

Acknowledgments

This paper was given at the workshop on Virtual Water in Agricultural Products: Quantification, Limitations and Trade Policy (Lincoln, Nebraska, USA, 14-16 September 2016), which was sponsored by the OECD Co-operative Research Programme: Biological Resource Management for Sustainable Agricultural Systems whose financial support made it possible for the second author to participate in the workshop.

Disclosure statement

No potential conflict of interest was reported by the authors.

ORCID

Mesfin M. Mekonnen http://orcid.org/0000-0002-3573-9759

References

Blas, A., Garrido, A., & Willaarts, A. B. (2016). Evaluating the water footprint of the mediterranean and American diets. *Water*, *8*(10). doi:10.3390/w8100448

Buzby, J., Hodan, W., & Jeffrey, H. (2014). *The estimated amount, value, and calories of postharvest food losses at the retail and consumer levels in the United States*. Washington, DC: US Department of Agriculture, Economic Research Service.

FAO, IFAD, & WFP. (2015). *The State of food insecurity in the World 2015. Meeting the 2015 international hunger targets: Taking stock of uneven progress*. Rome: Food and Agriculture Organization (FAO).

Foley, J. A., Ramankutty, N., Brauman, K. A., Cassidy, E. S., Gerber, J. S., Johnston, M., & Zaks, D. P. M. (2011). Solutions for a cultivated planet. *Nature*, *478*(7369), 337–342.

Gjerris, M., & Gaiani, S. (2013). Household food waste in Nordic countries: Estimations and ethical implications. *Etikk I Praksis - Nordic Journal of Applied Ethics*, *7* (1), 6–23.

Godfray, H. C. J., Beddington, J. R., Crute, I. R., Haddad, L., Lawrence, D., Muir, J. F., … Toulmin, C. (2010). Food Security: The Challenge of feeding 9 billion people. *Science*, *327* (5967), 812–818.

HHS & USDA. (2015). *2015–2020 dietary guidelines for Americans*. Washington, DC: U.S. Department of Health and Human Services (HHS) and U.S. Department of Agriculture (USDA). http://health.gov/dietaryguidelines/2015/guidelines/

Hoekstra, A. Y., Chapagain, A. K., Aldaya, M. M., & Mekonnen, M. M. (2011). *The water footprint assessment manual: Setting the global standard*. London, UK: Earthscan.

Jalava, M., Guillaume, J. H. A., Kummu, M., Porkka, M., Siebert, S., & Varis, O. (2016). Diet change and food loss reduction: What is their combined impact on global water use and scarcity? *Earth's Future*, *4*(3), 62–78.

Jalava, M., Kummu, M., Porkka, M., Siebert, S., & Varis, O. (2014). Diet change—A solution to reduce water use? *Environmental Research Letters*, *9*(7), 074016.

Kummu, M., de Moel, H., Porkka, M., Siebert, S., Varis, O., & Ward, P. J. (2012). Lost food, wasted resources: Global food supply chain losses and their impacts on freshwater, cropland, and fertiliser use. *Science of the Total Environment*, *438*, 477–489.

Meier, T., & Christen, O. (2013). Environmental impacts of dietary recommendations and dietary styles: Germany as an example. *Environmental Science & Technology*, *47*(2), 877–888.

Mekonnen, M., & Hoekstra, A. (2012). A global assessment of the water footprint of farm animal products. *Ecosystems, 15*(3), 401–415.

Mekonnen, M. M., & Hoekstra, A. Y. (2011a). *National water footprint accounts: The green, blue and grey water footprint of production and consumption* (Report No. 50). Delft. Retrieved from Value of Water Research Report Series.

Mekonnen, M. M., & Hoekstra, A. Y. (2011b). The green, blue and grey water footprint of crops and derived crop products. *Hydrology and Earth System Sciences, 15*(5), 1577–1600.

Tilman, D., Balzer, C., Hill, J., & Befort, B. L. (2011). Global food demand and the sustainable intensification of agriculture. *Proceedings of the National Academy of Sciences, 108*(50), 20260–20264.

Tilman, D., & Clark, M. (2014). Global diets link environmental sustainability and human health. *Nature, 515*(7528), 518–522.

Tom, M. S., Fischbeck, P. S., & Hendrickson, C. T. (2016). Energy use, blue water footprint, and greenhouse gas emissions for current food consumption patterns and dietary recommendations in the US. *Environment Systems and Decisions, 36*(1), 92–103.

USDA & HHS. (2010). *Dietary guidelines for Americans 2010* (7th ed.) U.S. Department of Agriculture (USDA) and U.S. Department of Health and Human Services (HHS). Washington, DC: U.S. Government Printing Office.

USDA-ERS. (2017, August21). Loss-Adjusted Food Availability (LAFA) data series. Available from U.S. Department of Agriculture - Economic Research Service (USDA-ERS) Retrieved from https://www.ers.usda.gov/data-products/food-availability-per-capita-data-system/

Vanham, D., Bouraoui, F., Leip, A., & Bidoglio, G. (2015). Lost water and nitrogen resources due to EU consumer food waste. *Environmental Research Letters, 10*(8), 084008.

Vanham, D., del Pozo, S., Pekcan, A. G., Keinan-Boker, L., Trichopoulou, A., & Gawlik, B. M. (2016). Water consumption related to different diets in mediterranean cities. *Science of the Total Environment, 573*, 96–105.

Vanham, D., Hoekstra, A. Y., & Bidoglio, G. (2013b). Potential water saving through changes in European diets. *Environment International, 61*, 45–56.

Vanham, D., Mekonnen, M. M., & Hoekstra, A. Y. (2013a). The water footprint of the EU for different diets. *Ecological Indicators, 32*, 1–8.

West, P. C., Gerber, J. S., Engstrom, P. M., Mueller, N. D., Brauman, K. A., Carlson, K. M., & Siebert, S. (2014). Leverage points for improving global food security and the environment. *Science, 345*(6194), 325–328.

Water footprint for Korean rice products and virtual water trade in a water-energy-food nexus

Sang-Hyun Lee, Jin-Yong Choi, Seung-Hwan Yoo and Rabi H. Mohtar

ABSTRACT

One possible approach for addressing water and food insecurity involves food production, trade, and water used elsewhere. In this study, we introduce a water footprint for Korean rice products and focus on the impacts of localized cultivation and water supply systems on the water footprint. In addition, we discuss several studies on the application of water footprint and virtual water trade in water and food management in Korea. Finally, we suggest the role of water footprint and virtual water trade in sustainable resource management through a water-energy-food nexus approach.

Introduction

Population growth and lifestyle changes affect the demand, supply, and trade of food. In particular, changes in food consumption pattern can affect food supply. These are strongly related to water resource management for food security, and water use for food production can be quantified through the concept of virtual water and water footprints. Virtual water indicates the amount of water use for producing goods and food trade can be regarded as embedded water trade in terms of virtual water which is termed virtual water trade (VWT). Water footprint includes virtual water concept and a product's entire value chain. Water footprint and VWT are relevant to national water policy and provide a new perspective for considering food and water security in sustainable resource management. Consideration should be given to relationships between water, food, and external factors, such as international trade and climate change.

In Korea, rice is a staple cereal crop that occupies approximately 60% of the farmland, and self-sufficiency rate for rice is almost 100%. In addition, paddy rice fields occupied 0.78 million ha in 2015, with over 80% of them irrigated (Lee, Yoo, Choi, & Bernard, 2016c). Therefore, the water footprint of rice is an important water-food indicator in the country. To estimate the water footprint of rice, firstly we needed to understand the cultivation system for paddy rice, which includes transplanting and

---

The opinions expressed and arguments employed in this publication are the sole responsibility of the authors and do not necessarily reflect those of the OECD or of the governments of its Member countries.

ponding systems. In addition, daily water balance based on pond water depth was applied to estimate paddy water use. Rice consumption is also an important issue in terms of food security and water management through the water footprint concept. In addition, Korea imports grains such as wheat, barley and maize. Crop products trade is related to domestic crop production, and is consequently related to water management and security in terms of VWT. Therefore, in this study, we showed the application of water footprint and VWT in water and food management in Korea. Furthermore, we suggested the role of water footprint and VWT in sustainable food and water management through a water-energy-food (WEF) nexus approach.

Water footprint of paddy rice in Korea using a water balance model

Water balance model in a paddy field under a ponding system

Crop water requirement is essential for estimating the water footprint of a crop, and it is dependent on the localized methods and parameters. For example, paddy rice is mainly cultivated in ponding systems in Korea, and irrigation water is supplied to maintain an optimal pond water depth (OPD) for each growth stage. In addition, a transplanting system is applied and a large volume of irrigation water is required during the transplanting season. Therefore, the water footprint of paddy rice in Korea is different from other countries' water footprint calculations.

Estimation of the water requirement for paddy rice in ponding systems was based on a daily water balance model including various factors, such as runoff, precipitation, irrigation, deep percolation, evapotranspiration, and pond water depth, as shown in Figure 1.

The runoff in ponding systems indicates the water released over the bund, and the depth of water is decreased by the water loss through crop evapotranspiration and percolation. Therefore, the net irrigation water requirement (NIWR) indicates the water supply for maintaining the OPD. The effective rainfall (ER) indicates the available rainfall that contributes to the pond water. Generally, ER indicates the amount of

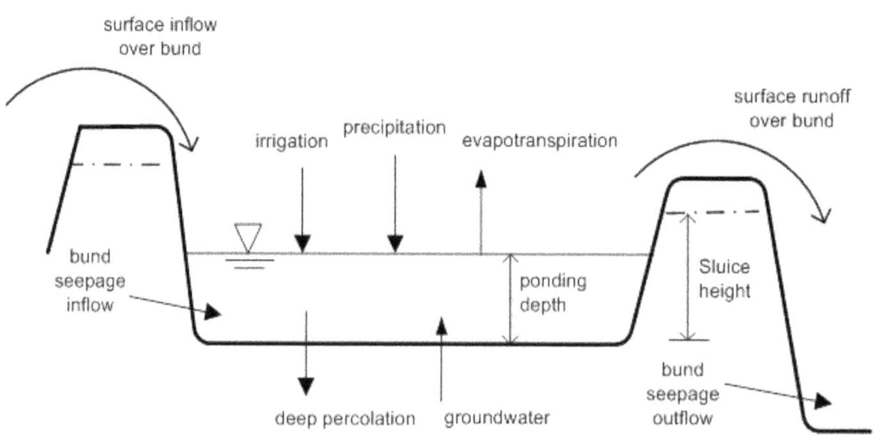

Figure 1. A ponding system in paddy rice fields.

rainfall that contributes to water captured by soil, however; in paddy field by ponding system soil is always saturated and rainfall is directly involved into pond water, and ER carries over in water balance model. The water balance model is described as shown in Equation 1 (Jensen, Burman, & Allen, 1990), and the NIWR and ER in the water balance model are described as shown in Equation 2:

$$PD_t = PD_{t-1} - ET_{ct} - DP_t + (RF_t - SR_t) \tag{1}$$

$$NIWR_t = OPD_t - PD_t \text{ for } OPD_t > PD_t \tag{2}$$

$NIWR_t = 0$ for $OPD_t < PD_t$
$ER_t = RF_t$ for $SR_t = 0$,
$ER_t = RF_t - SR_t$ for $SR_t > 0$

where t is time (d), PD is depth of pond water (mm), OPD is the optimal depth of pond water (mm), ET_c is crop evapotranspiration, SR is surface runoff in the paddy field outlet (mm), DP is deep percolation (mm), RF is rainfall (mm), NIWR is net irrigation water requirement (mm), and ER is effective rainfall.

Most paddy rice irrigation systems were developed based on Korea's relatively homogenous soil, which is clay textured and has relatively low percolation rates (Jang, Choi, & Lee, 2007). The Rural Research Institute in Korea suggests that 5.0–5.6 mm/d of water deep percolates into paddy fields based on field experiments conducted in several basins since 1970 (Korea Water Resources Cooperation [KWRC], 2002). In addition, previous studies showed that about 4–6 mm/d of water percolates into paddy soils during the irrigation season (Lee, 1988). Yoo, Choi, Lee, and Kim (2014) applied a deep percolation value of 5.0 mm/d in water footprint accounts for paddy rice.

Evapotranspiration and crop coefficient for paddy rice

Reference evapotranspiration (ET_0) is an important factor for evapotranspiration of each crop, and plays a key role in estimating crop growth, water demand, and crop water management (Nam, Hong, & Choi, 2015). Changes in ET_0 are of great significance for water resource planning, irrigation requirements, and agricultural production (Hatch, Jagtap, Jones, & Lamb, 1999; Peterson & Keller, 1990). The ET_0 was calculated using the Penman-Monteith equation as derived by Allen, Pereira, Raes, and Smith (1998), which allows the user to input necessary daily, weekly, and monthly atmospheric weather data (air temperature, humidity, radiation, and wind speed). Equation 3 is as follows (Allen et al., 1998):

$$ET_0 = \frac{0.408\Delta(R_n - G) + \gamma(900/T + 273)u_2(e_s - e_a)}{\Delta + \gamma(1 + 0.34u_2)} \tag{3}$$

where ET_0 is the reference crop evapotranspiration (mm/d), ET_c is the crop evapotranspiration (mm/d), K_c is the crop coefficient, Δ is the slope of saturated vapor pressure/temperature curve (kPa/°C), γ is the psychrometric constant (kPa/°C), u_2 is the wind speed at 2 m height (m/s), R_n is the total net radiation at the crop surface (MJ/(m^2/d)), G is the soil heat flux density (MJ/(m^2/d)), T is the mean daily air temperature at 2 m height (°C), e_s is the saturation vapor pressure (kPa), and e_a is the actual vapor pressure (kPa).

The ET_o can be multiplied by the crop coefficient (K_c), which is a unique factor that takes into account the plant properties, to determine crop evapotranspiration (ET_c), as suggested by Equation 4.

$$ET_c = ET_0 \times K_c \tag{4}$$

where ET_o is the reference crop evapotranspiration (mm/d), ET_c is the crop evapotranspiration (mm/d), and K_c is the crop coefficient.

K_c can be influenced by cultivation, local climatic conditions, and seasonal differences in crop growth patterns (Kuo, Ho, & Liu, 2006). Accordingly, the localized K_c should be applied to calculate the ET_c and water footprint. In Korea, Yoo, Choi, and Jang (2008) estimated K_c using observed and simulated ET_c, as shown in Figure 2.

Optimal pond water depth for paddy fields

In paddy fields with a ponding system, understanding paddy rice growth stages was required to understand the OPD, which is the main factor that determines irrigation water requirements. First, transplanting water should be supplied to keep paddy fields saturated and to set the initial pond water depth. Yoo et al. (2008) and Lee et al. (2016c) set the transplanting water demand in Korea to total 140 mm during the transplanting season (generally around 30 days), as suggested by Ministry of Agriculture and Forestry (MAF) (1998). One month after the transplanting season, midsummer drainage is performed to increase production. After drainage, the OPD increases to 40 mm, as suggested by Doorenbos and Kassam (1979) and Jang, Choi, Park, Bae, and Chung (2004) (Figure 3). To conclude, a paddy field outlet height should be set for estimating runoff. Then, the required irrigation water is supplied to maintain the OPD in each growth stage.

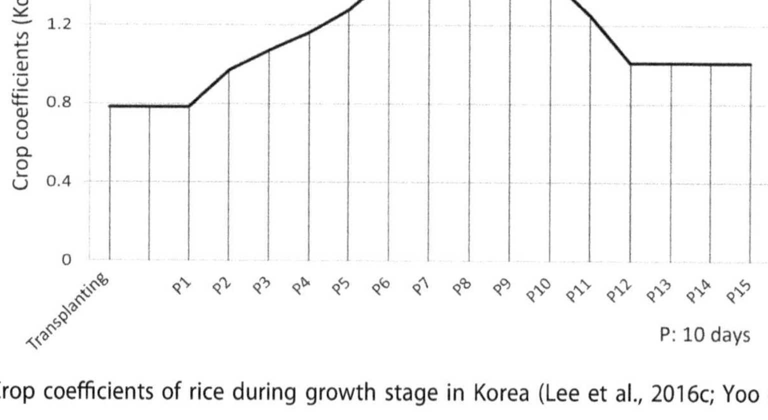

Figure 2. Crop coefficients of rice during growth stage in Korea (Lee et al., 2016c; Yoo et al., 2008).

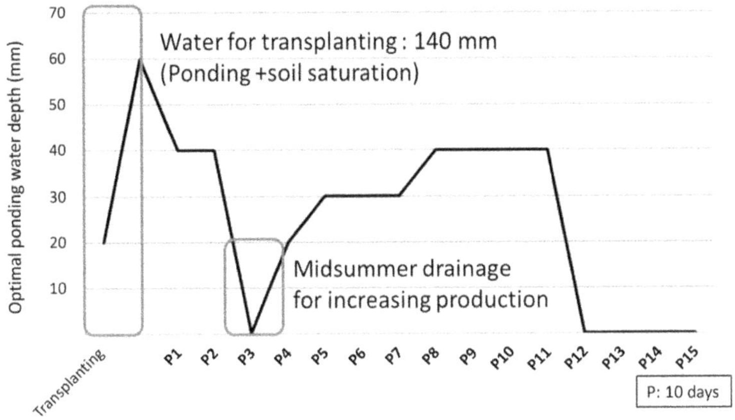

Figure 3. Optimal pond water depth in paddy fields based on growth stages of rice (Jang et al., 2004; Yoo et al., 2008; Lee et al., 2016c).

Green and blue water footprints of paddy rice

Green water footprint generally indicates rainfall captured in the soil. However, in Korean paddy ponding systems, green water represents the rainfall that contributes to increased pond water depth, with the exception of loss due to surface runoff overflow through paddy field outlets; this is termed ER. In the water balance model, the green water footprint is calculated by summing the ER during growth periods. Blue water indicates irrigation water supplied from an artificial facility, which is generally a reservoir. In Korean paddy ponding systems, blue water indicates the NIWR. Finally, green and blue water footprint is calculated as in Equation 5:

$$\mathrm{GWF} = \frac{10 \times \sum \mathrm{ER_t(mm/day)} \times \mathrm{Paddy\,area(ha)}}{\mathrm{Production(ton)}} \tag{5}$$

$$\mathrm{BWF} = \frac{10 \times \sum \mathrm{NIWR_t(mm/day)} \times \mathrm{Paddy\,area(ha)}}{\mathrm{Production(ton)}}$$

where ER is effective rainfall (mm/day), NIWR is net irrigation water requirement (mm/day), Paddy area is the area of paddy fields (ha), Production is the production of rice (ton), GWF is green water footprint (m^3/ton), and BWF is blue water footprint (m^3/ton).

Table 1 shows the different methods for estimating water footprint of paddy rice used by Yoo et al. (2014) and Chapagain and Hoekstra (2010). The main difference between the two studies is the cultivation type. In Korea, paddy rice is cultivated in ponding systems, which indicates that the rice grows in saturated water. Therefore, soil moisture is not the main variable in ponding systems. Irrigation water and ER are estimated by water balance, which includes the water requirement for meeting optimal water depth in the paddy pond. Yoo et al. (2014) estimated the green and blue water footprints of paddy rice in Korea using a daily water balance model that considered the ponding system and localized values.

Table 1. Green and blue water footprint of paddy rice in Korea according to different estimation methods.

Items	Chapagain and Hoekstra (2010)	Yoo et al. (2014)
Meteorological Station	3 locations	69 locations
Period	2000–2004	2004–2009
Cultivation Type	Wetland system	Ponding system
Growing Period	120 days	120 days
Crop Water Requirement	CROPWAT4	Daily water balance
Evapotranspiration	FAO Penman-Monteith	FAO Penman-Monteith
Effective Rainfall	USDA SCS method	Freeboard model
Percolation	5.0 mm/d	4.0 mm/d
Transplanting Water	100 mm	140 mm
Time Step	5 days	1 day
Water Footprint (m^3/ton)	745	787
Green Water Footprint (m^3/ton)	356	295
Blue Water Footprint (m^3/ton)	389	502

Water footprint of paddy rice products based on a value chain in Korea

The water footprint of a final product needs to include the sub-processes for the final product. For example, rice is a primary crop for rice bran and rice flour. Therefore, the rice water footprint is used to determine that of processed products based on product and value fractions (Chapagain & Hoekstra, 2004; Hoekstra, Chapagain, Aldaya, & Mekonnen, 2011). The water footprint of a processed product is expressed using Equation 6:

$$WF_{Product}[p] = \left(\frac{WF_{Product}[i]}{p_f[p,i]}\right) \times v_f[p] \tag{6}$$

where $WF_{product}$ is the water footprint of a product, p_f is the product fraction, v_f is the value fraction, p is the processed product, and i is the input product or root product.

The product fraction (p_f) is defined as the weight of the processed product obtained from one ton of root product, and the value fraction (v_f) is the ratio between the market value of the processed product and that of all the processed products obtained from the root product (Chapagain & Hoekstra, 2010). If more than two processed products are obtained from a root product, it is necessary to share the water footprint of the root product over its processed products based on p_f and v_f (Yoo et al., 2014).

Figure 4 shows that the water footprint is based on the value chain from paddy rice to each rice product. For example, paddy rice can produce husked rice or brown rice. Based on survey data in Korea, 0.8 tons of brown rice are produced from 1 ton of paddy rice. Therefore, the p_f is 0.8. The v_f is related to the cost of paddy rice and brown rice. The price of brown rice is slightly higher than that of paddy rice; therefore, the v_f is 0.95 according to the method of Yoo et al. (2014). Based on p_f and v_f, the water footprint of brown rice is 1001 m^3/ton. Following the same process, the water footprint of white rice in Korea is 1055 m^3/ton. Since the v_f is based on the price of rice products, the water footprint is flexible.

Climate change impacts on water footprints in Korea

A water footprint is a localized value related not only to climate, but also to cultivation, irrigation, and water supply systems; pond water systems in Korean paddy field are a

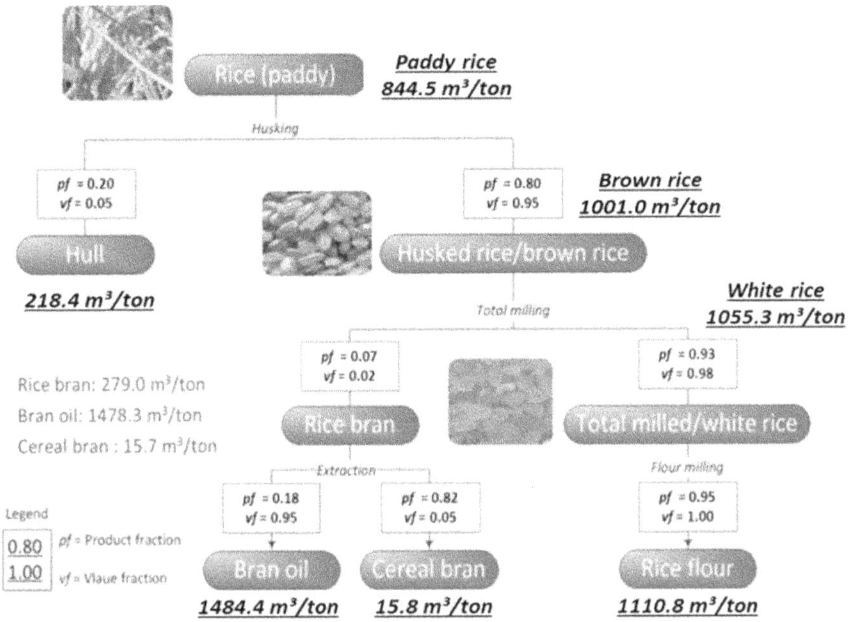

Figure 4. Water footprint of rice products based on product and value fraction in supply chain (Yoo et al., 2014a).

good example of this. In addition, the water footprint is strongly affected by climate change. Nam et al. (2015) depicted the spatial distributions of the annual ETo change and difference values between 1973–1992 and 1993–2012. They found that the southeast regions of the country had the highest values. Over the past 20 years (1993–2012), the spatial distributions of the difference values with annual ETo increased by approximately 30 mm around the northwest and southern areas, while decreasing around the northern, mid-western, and central areas, compared with the past values (1973–1992).

Irrigation water requirement is also dependent on crop phenology. In particular, dibbling rice seedlings in a ponding system involves intensive water use during the transplanting season. However, the transplanting season includes the drought season in Korea. Therefore, climate changes during the transplanting season can affect green and blue water requirements (Lee et al., 2016c). In addition, total irrigation water is strongly related to the length of the growing period. For example, a shorter growing period leads to a smaller amount of irrigation water use. In other words, the changes in rice phenology due to climate change affect the available rainfall and irrigation water requirements that are directly related to green and blue water footprints.

Lee et al. (2016c) analyzed the optimal transplanting date and growing period under temperature increases due to climate change. The rice phenology process was based on the accumulated temperature and heading date, as shown in Figure 5 (Lee et al., 2016c; Park & Lee, 2005; Yoo, Choi, Lee, Oh, & Yun, 2013). The main factor that determines the optimal transplanting date is the optimum ripening period. When the daily mean temperature is 21–23°C for 40 days after flowering, the heading date indicates the first

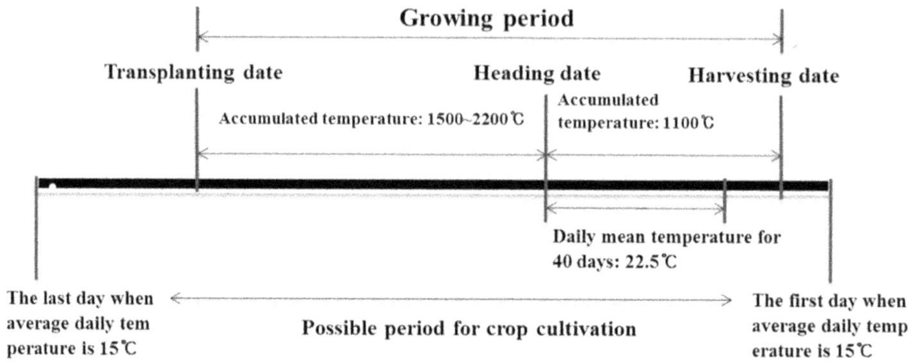

Figure 5. The process of optimal paddy rice phenology based on accumulated temperature (Lee et al., 2016c; Park & Lee, 2005; Yoo et al., 2013).

day of the optimum ripening period. Based on the heading date, the optimal transplanting date is determined by an inverse count of accumulated temperature from the heading date to the transplanting date (extremely early maturing rice: 1500°C; late maturing rice: 2200°C). The rice harvest starts on the date at which the accumulated temperature from the heading date reaches 1100°C.

Figure 6 shows the optimal transplanting date for 30 years from 1981 to 2100 (Lee et al., 2016c) and indicates that climate change influences the transplanting date due to increases in temperature. The average optimal transplanting date was determined to be June 5 for the 2025 period under the RCP (representative concentration pathway) 4.5 scenario, which is used to create projections based on the factors of greenhouse gas emissions and atmospheric concentrations, air-pollutant emissions, and land use (IPCC, 2014). Compared to the 1995 period, the optimal transplanting date was delayed by 25 days for the 2025 period. This date was pushed back to June 29 for the 2085 period (Lee et al., 2016c). Growing periods affected by climate change are projected to be shorter in future seasons due to increasing overall temperatures.

Changes in the transplanting season and growing period can strongly affect rainfall and irrigation water. Figure 7 shows the ER (green water) and net irrigation water (blue water) with the optimal transplanting and growing season compared with those in unchanging seasons in paddy rice fields (Lee et al., 2016c). Changing the optimal transplanting season in paddy rice can lead to increased availability of ER. The representative drought season in Korea is from March to May, which overlaps with the transplanting season. However, the increase in temperature due to climate change could delay the transplanting season until June, which is the early wet season. In addition, decreasing the growing period decreases total irrigation water. Climate change could cause more frequent severe droughts, but appropriate changes in the transplanting season and growing period due to climate change could also increase green water efficiency and blue water savings in paddy fields. Therefore, green and blue water footprints of rice in the future should consider climate change impacts related to both drought severity and changes in phenology.

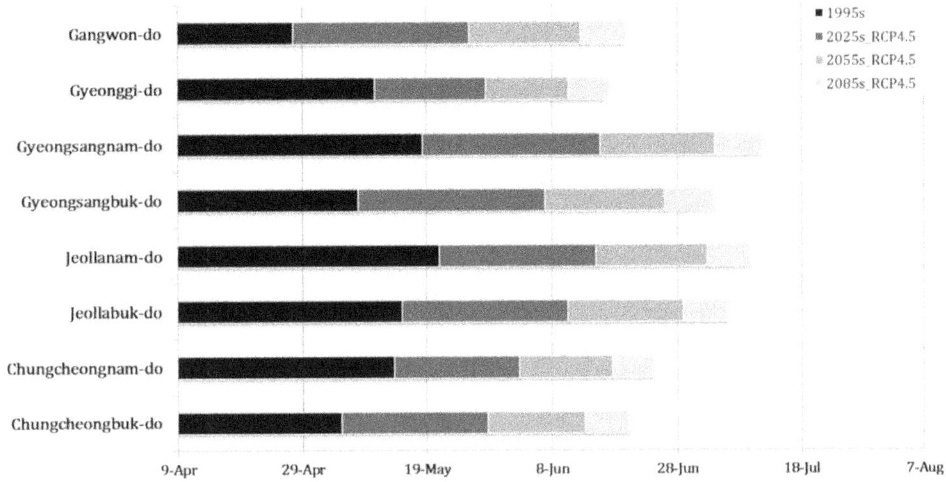

(a) Optimal transplanting date

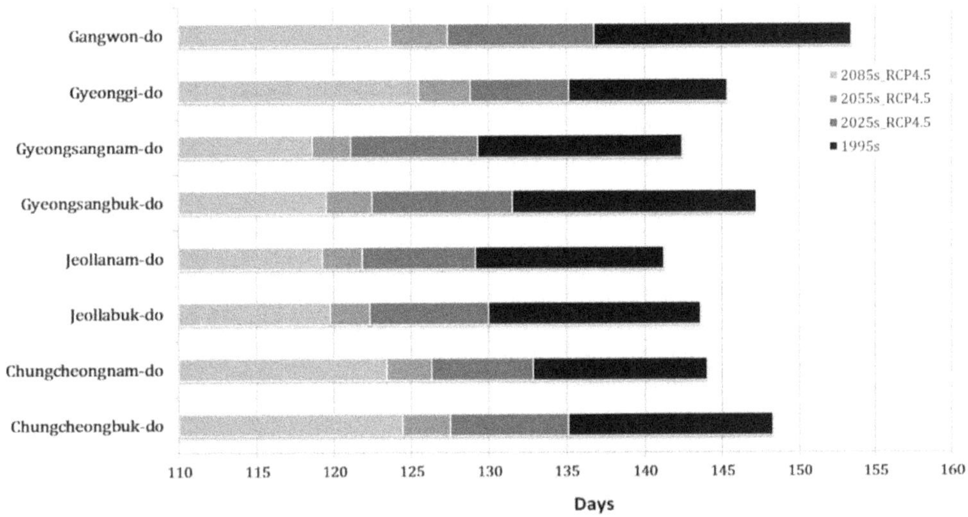

(b) Optimal growing period

Figure 6. The optimal transplanting date and growing period from 1995 to 2085 periods under RCP 4.5 scenario (Lee et al., 2016c).

Application of water footprint and virtual water

Virtual water use for Korean food security

The food self-sufficiency ratio (SSR) indicates the amount of domestic production in total consumption. In Korea, the food SSR decreased significantly from 56% in 1980 to 27% in 2010. This was due to changes in food consumption patterns and increases in food imports. Accordingly, the Korean government recognized the necessity to increase the food SSR. It set targets for the grain SSR at 32% in 2020 on the basis of the 'Report on the conceptualization of food SSR and adjustment of its target in Korea' (Korea

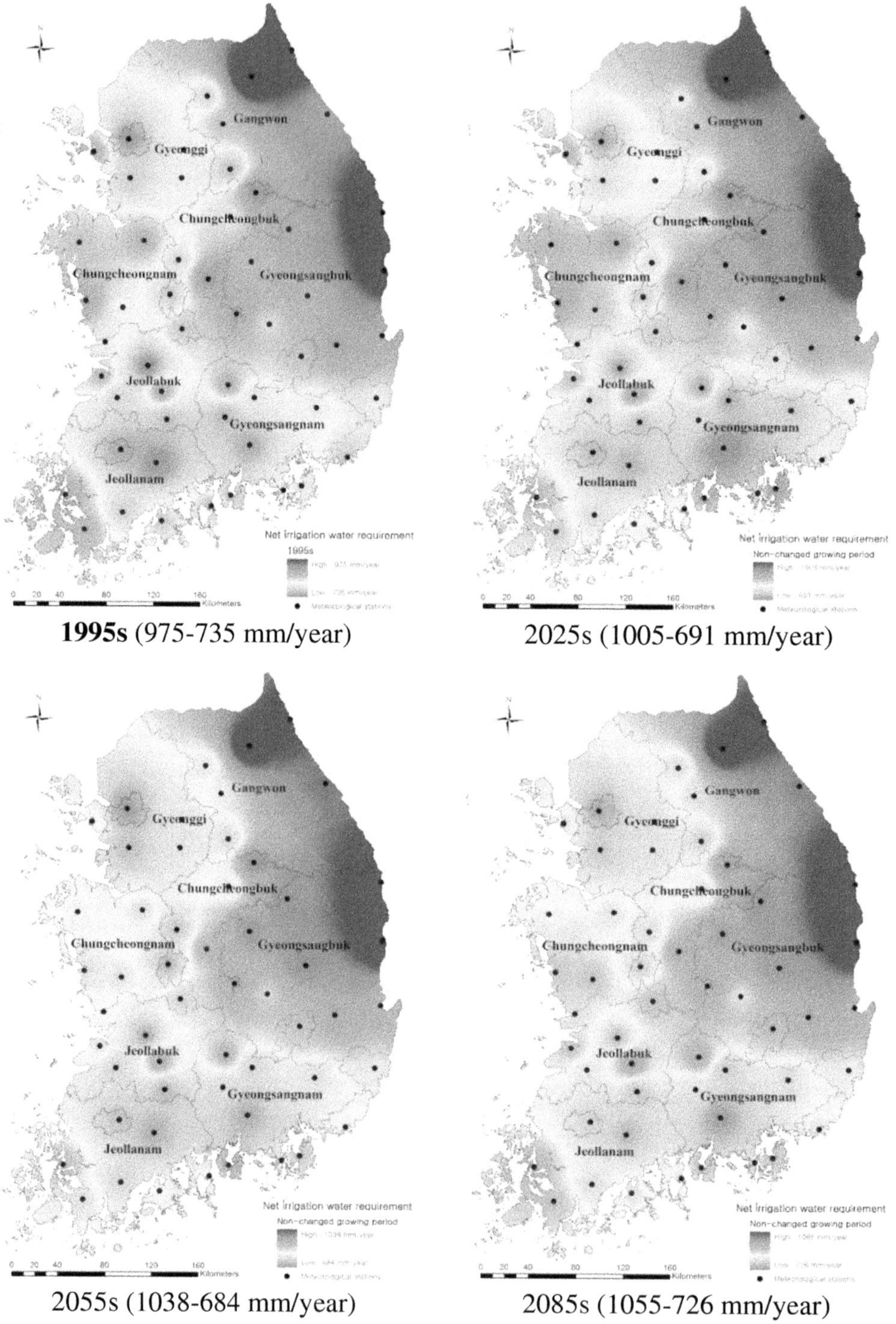

Figure 7. The distributed net irrigation water requirement under the RCP 4.5 scenario during the non-changed growing periods (Lee et al., 2016c).

Rural Economic Institute [KREI] and Ministry for Food, Agriculture, Forestry and Fisheries [MIFAFF], 2011). This report presents an analysis of the present situation and future outlook of the food SSR and possible food consumption and production development scenarios to establish food SSR targets in 2020. However, increasing food security requires additional water to increase food production. It is important to consider the trade-off between additional water use and increased food security. Water footprint can be used to estimate the additional water requirement for food security.

Traditional agricultural water management in Korea has generally focused on operating water resources, such as reservoirs and ground water, for a steady supply of food. However, water use for food production is related to the pattern of food consumption. For example, meat consumption has been increasing in Korea, and this diet change from cereals to meat causes more water use because the meat water footprint is much larger than that of the crop water footprint in Korea. In a study conducted by Lee, Choi, and Yoo (2015), virtual water use (m^3/yr) by food consumption per capita was analyzed using food supply data and the water footprint in Korea. Table 2 shows that about 330 m^3/yr of virtual water was consumed for main food consumption per capita in 1985, and it increased to 450 m^3/yr per capita in 2010. In particular, the rate of virtual water consumption for meat production increased from 28% in 1985 to 54% in 2010. Food consumption patterns have a significant impact on crop water requirements, cultivation area, and food self-sufficiency.

Virtual water trade in Korea

Virtual water also provides a new perspective on international trade regarding water resources (Novo, Garrido, & Varela-Ortega, 2009). VWT indicates the embedded water in a food trade; it is calculated by multiplying the food trade by the associated water footprint, as shown in Equation 7. Therefore, VWT can be regarded as the primary source of water demand in importing countries (Hoekstra, 2003).

Table 2. Virtual water consumption by food consumption from 1985 to 2010 (Lee et al., 2015a).

		Virtual water consumption by food consumption (m^3/cap/yr)					
Foods		1985	1990	1995	2000	2005	2010
Cereals	Rice	127.4	120.2	110.0	97.3	82.8	81.0
	Wheat	33.9	31.5	36.1	38.3	33.6	35.3
	Barely	6.7	1.9	1.5	1.5	0.9	1.1
	Maize	17.3	23.2	25.4	30.0	32.9	27.4
	Others	0.8	0.6	4.8	5.2	6.6	6.0
Meat	Bovine	49.7	70.2	113.1	141.7	108.1	143.5
	Pig	35.6	49.9	60.3	67.4	71.6	78.2
	Poultry	7.4	9.7	10.3	12.0	14.1	20.0
Pulses	Soy beans	30.9	27.7	30.2	28.5	30.7	28.2
	Red beans	3.2	3.1	2.5	2.3	2.2	2.0
	Others	0.8	0.6	4.8	5.2	6.6	6.0
Starchy roots	potatoes	0.9	0.9	1.0	1.0	1.7	1.2
	Sweet potatoes	2.0	1.7	1.4	1.7	1.7	1.8
Vegetables		13.5	18.1	22.0	22.7	19.9	18.1
Fruits		11.5	12.5	16.9	17.5	19.3	19.0
Total		**341.5**	**371.9**	**440.2**	**472.2**	**432.6**	**469.1**

$$VWT[n_e, n_i, c, t] = CT[n_e, n_i, c, t] \times WF[n_e, c] \quad (7)$$

where VWT denotes the VWT from the exporting country (n_e) to the importing country (n_i) in year t as a result of trade in crop c. CT represents the crop trade from the exporting country n_e to the importing country n_i in year t as a result of trade in crop c. WF represents the water footprint of crop c in the exporting country.

Approximately 13% of the water used for worldwide crop production was traded internationally from 1995–1999 (Hoekstra & Hung, 2005), and 16% of global water resources were used in production and export (Chapagain & Hoekstra, 2008). Korea is one of the major countries that import food and feed crops. Significant increases in domestic food production will be required for food security in situations of food import failure. Consequently, significant water requirements should be implemented for food trade in Korea as a key factor for sustainable water management. Lee, Yoo, Shin, and Choi (2016b) calculated the VWT values of 41 crops in Korea from 2006 to 2010. A volume of 96.2 Gm^3 of virtual water was imported via crop trade, and the most virtual water was imported through wheat (28.4 Gm^3), as shown in Table 3.

Through the virtual water concept, the internal and external water footprints can be analyzed, which is the virtual water budget according to products, trade, and consumption of commodities (Chapagain & Hoekstra, 2004). The internal water footprint indicates the total water used for food production, and the external water footprint indicates foreign water imported through food trade. If the external water footprint of a country represents a major portion of the total water use, including domestic water and virtual water import, then the country has a high level of dependency on foreign water resources, which indicates that virtual water is imported. Lee et al. (2016b) showed that Korea has a high level of dependency on foreign water resources through virtual water import, except for rice. We analyzed the water dependency on foreign water resources using the internal water footprint and the external water footprint. The internal water footprint of wheat was 86 Mm^3; in contrast, the external water footprint was 28,436 Mm^3 in Korea, as shown in Table 4. Total internal and external water footprint of main crops were 39,513 and 96,186 Mm^3, respectively. This means that the virtual water imported through crop trade is an important parameter not only in terms of food security, but also in terms of water security in Korea.

Table 3. Virtual water trade by crop trade in Korea during the period 2006–2010 (Lee et al., 2016b).

	Net Import from 2006 to 2010			
Crops	Crop (1000 ton)	Green Water (Mm^3)	Blue Water (Mm^3)	Total Water (Mm^3)
Wheat	16,000	27,000	1,430	28,400
Rice	1,494	1,438	888	2,334
Barley	166	203	10	213
Others *	168	255	8	263
Root & tuber	89	8	7	16
Maize	45,100	29,600	2,710	32,300
Pulse crops	15,487	31,104	675	31,804
Vegetables	820	87	6	93
Fruits	3	−70	14	−56
Total	79,441	89,595	5,754	95,389

*rye and sorghum

Table 4. Estimation of water dependency using virtual water trade in Korea during the period 2006–2010 (Lee et al., 2016b).

		Internal water footprint (Mm^3)			External water footprint (Mm^3)		
Crops		Green Water	Blue Water	Total Water	Green Water	Blue Water	Total Water
Grain Crops	Wheat	86	0	86	27,004	1433	28,436
	Rice	9495	15,533	25,028	1444	892	2336
	Barley	807	0	807	203	10	213
	Others*	150	0	150	255	8	263
Root and Tuber Crops		1008	0	1008	8	7	16
Maize		408	0	408	29,564	2710	32,274
Pulse Crops		1840	0	1840	31,729	675	32,403
Vegetables		5254	666	5921	187	8	195
Fruits		4264	0	4264	37	14	51
Total		23,314	16,199	39,513	90,430	5756	96,186

*Rye and sorghum

The roles of water footprints in the water-energy-food nexus

Resource security has been threatened by climate change, population increase, and economic situations. In the 2015 World Economic Forum report, water, food crises, and energy shock were regarded as the main future concerns. In addition, the United Nations has established 17 Sustainable Development Goals. One of these is integrated resource management. With regard to water, these goals call for increased water usage efficiency across all sectors, as well as implementation of integrated water resource management. This will also impact the need for increased food production with environmentally sustainable resource management. The relationship between water, food, and trade can be regarded as the key factor in the WEF nexus system for assessing various strategies and external impacts from a holistic viewpoint, and for supporting decision makers by providing evaluation information on the different implications of current decisions on projected pathways.

In typical resource management, agricultural water management considers the water supply to the crop area. However, the WEF nexus includes virtual water, food demand according to population increase, international food trade, and integrated management of water and food, which can be assessed through this nexus. In particular, VWT could be used to assess the impacts of food trade on water security because the replacement of crop imports by domestic production in importing countries will accompany additional water use (Lee, Mohtar, Choi, & Yoo, 2016a). In addition, policy makers and resource managers can consider dynamic water allocation among agriculture, hydroelectricity, and municipal areas through the nexus system. Therefore, the relationship between water and food, and the impacts of food trade on the virtual water requirements introduced in this study, are fundamental information for the development of a WEF nexus system.

For example, the WEF nexus system can be a guide to assess the impacts of global food trade on water, land, and energy savings. As international grain crop trade expands with increased connections, the changes in trade could affect global and national food security. In addition, the water saving effects of food trade could be assessed using a virtual water concept that indicates the embedded water for the produced commodity. Food trade also affects land, energy, and carbon emissions. National water, land, and energy savings in each importing country can be quantified when shifting from import to domestic production. National energy savings are indicated by the amount of electricity consumed for blue water supply to crops. The national energy savings consider the type of water resources, such as ground water, surface water, and treated water, because each country has a different distribution of water resources. For example, the same amount of water supply could derive different energy savings depending on the type of water resource because treated water consumes more energy than surface water. In addition, energy savings are related to carbon emission savings. We quantified the reduction in carbon emissions by energy savings. Therefore, the WEF nexus system considers trade-offs between food security, resource savings, and carbon emissions. These findings could contribute to the optimization of trade strategies.

Conclusions

One possible approach for addressing water and food insecurity involves crop imports and water used elsewhere through the concept of water footprint and VWT. Water is also often consumed with little regard for its true value. Therefore, when considering crops to grow versus crops to import, one needs to consider the real value of water in crop production. Consequently, we intend to develop water value measurements based on shortage costs and current water development activities. These can then be integrated with crop production budgets and water duties to obtain the total cost of crop production when comparing international trade possibilities. This may reveal crops that would be more advantageous to import in scarce water environments. It may also make water conserving irrigation and water conveyance methods more desirable and help determine which conservation practices to promote.

However, there are also limitations when applying water footprint and VWT to water management. First, applying water footprints to food consumption introduces new concepts of water management, but it is also difficult to link to operating water resource systems, such as reservoirs because water footprint indicates sum of water used during growing period but weekly or monthly operation plans could be set by a reservoir manger. However, water footprint can be useful for long-term or national plan of water resource management. For example, potential water use by the changes in food consumption pattern and population could be analyzed by water footprints and it could be useful information for managing long-term operation and capacity of water resource. Second, VWT is calculated using water footprint and food trade, and their time scales can be different. However, this limitation can be overcome by regular update of water footprint. In addition, the water footprint acts as an index for quantifying water requirements for food, and the average value of water footprint during significant period could be adopted as a consistent index. Finally, trade policy considers economic value based on the price of food, but there is no global value established for virtual water. Now water security has been significantly important issue and water could be main driver in food trade. In addition, food price could be affected by water scarcity in exporting countries. Therefore, virtual water trade could give new insight into food trade in the situation of global water shortage.

Even with the limitations of water footprints and VWT, they still provide information on technical and social aspects that is required to integrate sustainable water resource management with food security. Follow-up research to analyze water requirements for food consumption and food self-sufficiency and to evaluate VWT impacts on water and food security should be conducted with particular emphasis on sustainable agricultural production and water resource management in Water-Energy-Food Nexus context.

Acknowledgments

The workshop "Virtual Water in Agricultural Products: Quantification, Limitations and Trade Policy" took place in Lincoln, Nebraska, USA, September 14–16, 2016. The Workshop was sponsored by the OECD Co-operative Research Programme: Biological Resource Management for Sustainable Agricultural Systems, whose financial support made it possible for most of/some of the invited speakers to participate in the Workshop. The conference organizer was the University of Nebraska. One of the invited expert speakers at this conference was Sanghyun Lee of Texas A&M University. His presentation was entitled "Water footprint for rice products in Korea and virtual water trade in Water-Energy-Food Nexus.

Disclosure statement

No potential conflict of interest was reported by the author.

References

Allen, R. G., Pereira, L. S., Raes, D., & Smith, M. (1998). Crop evapotranspiration: Guidelines for computing crop requirements. FAO Irrigation and Drainage, Paper No. 56. Rome, Italy: Food and Agriculture Organization of the United Nations.

Chapagain, A. K., & Hoekstra, A. Y., (2004). *Water footprints of nations* (Value of Water Research Report Series No. 16). Delft, Netherlands: UNESCO-IHE.

Chapagain, A. K., & Hoekstra, A. Y. (2008). The global component of freshwater demand and supply: An assessment of virtual water flows between nations as a result of trade in agricultural and industrial products. *Water International*, *33*(1), 19–32.

Chapagain, A. K., & Hoekstra, A. Y., (2010). *The green, blue and grey water footprint of rice from both a production and consumption perspective* (Value of Water Research Report Series No. 40). Delft, Netherlands: UNESCO-IHE Institute for Water Education.

Doorenbos, J., & Kassam, A. H., (1979). Yield response to water. FAO Irrigation and drainage Paper No. 33. Food and Agriculture Organization of the United Nations, Rome.

Hatch, U., Jagtap, S., Jones, J., & Lamb, M. (1999). Potential effects of climate change on agricultural water use in the southeast U.S. *Journal of the American Water Resources Association*, *35*(6), 1551–1561.

Hoekstra, A. Y., (2003). *Virtual water trade: Proceedings of the international expert meeting on virtual water trade* (Value of Water Research Report Series No. 12). Delft, Netherlands: UNESCO-IHE.

Hoekstra, A. Y., Chapagain, A. K., Aldaya, M. M., & Mekonnen, M. M. (2011). *The water footprint assessment manual: Setting the global standard*. London: Earthscan Publishing.

Hoekstra, A. Y., & Hung, P. Q. (2005). Globalisation of water resources: International virtual water flows in relation to crop trade. *Global Environmental Change*, *15*(1), 45–56.

Intergovernmental Panel on Climate Change (IPCC). (2014). *Climate change 2014 – Impacts, adaptation and vulnerability: Regional aspects*. Cambridge University Press: Cambridge, UK.

Jang, M. W., Choi, J. Y., & Lee, J. J. (2007). A spatial reasoning approach to estimating paddy rice water demand in Hwanghaenam-do, North Korea. *Agric Water Manag*, *89*, 185–198.

Jang, M. W., Choi, J. Y., Park, K. W., Bae, S. J., & Chung, H. W. (2004). Development of a single reservoir agricultural drought evaluation model for paddy. *Journal Korean Social Agricultural Engineering, 46*(3), 3–17 in Korean, with English Abstract.

Jensen, M. E., Burman, R. D., & Allen, R. G. (1990). Evapotranspiration and irrigation water requirements. *ASCE, 70*, 332.

Korea Rural Economic Institute (KREI), Ministry for Food, Agriculture, Forestry and Fisheries (MIFAFF). (2011). *Report on the conceptualization of food self-sufficiency ratio and adjustment of its target in Korea.* Seoul, Republic of Korea (in Korean): Korea Rural Economic Institute.

Korea Water Resources Cooperation (KWRC). (2002). Dam design manual (in Korean). Seoul, Republic of Korea: KWRC. doi:10.1044/1059-0889(2002/er01)

Kuo, S. F., Ho, S. S., & Liu, C. W. (2006). Estimation irrigation water requirements with derived crop coefficients for upland and paddy crops in ChiaNan Irrigation Association, Taiwan. *Agriculture Water Management, 82*(3), 433–451.

Lee, N. H., (1988). *Simulating daily operation of water management system of irrigation districts* (Ph.D. thesis). Seoul National University, Seoul, Republic of Korea (in Korean, with English Abstract). doi:10.3168/jds.S0022-0302(88)79586-7

Lee, S. H., Choi, J. Y., & Yoo, S. H. (2015). Estimation of the virtual water consumption for food consumption and calorie supply. *Journal of the Korean Society of Agricultural Engineers (KSAE), 57*(3), 77–86 (in Korean).

Lee, S. H., Mohtar, R. H., Choi, J. Y., & Yoo, S. H. (2016a). Analysis of the global agricultural virtual water trade network using degree and eigenvector centrality, hydrol. *Earth Systems Sciences, 20*, 4223–4235.

Lee, S. H., Yoo, S. H., Choi, J. Y., & Bernard, A. E. (2016c). Effects of climate change on paddy water use efficiency with temporal change in the transplanting and growing season in Korea. *Irrigation and Science, 34*(6), 443–463.

Lee, S. H., Yoo, S. H., Shin, A., & Choi, J. Y. (2016b). Evaluation of the dependency and intensity of virtual water trade in Korea. *Irrigation and Drainage, 65*, 48–56.

Ministry of Agriculture and Forestry (MAF). (1998). *Design criteria of land and water development plan for agriculture (Irrigation).* Gwachon: Ministry of Agriculture and Forestry (in Korean).

Nam, W. H., Hong, E. M., & Choi, J. Y. (2015). Has climate change already affected the spatial distribution and temporal trends of reference evapotranspiration in Korea? *Agriculture Water Manage, 150*, 129–138.

Novo, P., Garrido, A., & Varela-Ortega, C. (2009). Are virtual water "flows" in Spanish grain trade consistent with relative water scarcity? *Ecological Economics, 68*, 1454–1464.

Park, S. J., & Lee, J. H. (2005). *Edible plants.* Seoul (in Korean): Korea National Open University Press.

Peterson, D. F., & Keller, A. A. (1990). Effects of climate change on U.S. irrigation. *Journal of Irrigation and Drainage Engineering, 116*(2), 194–210.

Yoo, S. H., Choi, J. Y., & Jang, M. W. (2008). Estimation of design water requirement using FAO Penman-Monteith and optimal probability distribution function in Korea. *Agric Water Manag, 95*, 845–853.

Yoo, S. H., Choi, J. Y., Lee, S. H., & Kim, T. (2014). Estimating water footprint of paddy rice in Korea. *Paddy and Water Environment (PAWE), 12*, 43–54.

Yoo, S. H., Choi, J. Y., Lee, S. H., Oh, Y. G., & Yun, D. K. (2013). Climate change impacts on water storage requirements of an agricultural reservoir considering changes in land use and rice growing season in Korea. *Agriculture Water Management, 117*, 43–54.

Water footprint of beef production on Texas High Plains pasture

Charles P. West and Lisa L. Baxter

ABSTRACT

Depletion of the Ogallala aquifer in the U.S. Great Plains demands information on water usage in forage-fed beef production. We compared a summer grazing system for beef steers consisting of warm-season grasses vs. grasses growing with lucerne (*Medicago sativa* L.). The water footprint averaged 39 and 26 m^3 of water kg^{-1} of beef liveweight gain, respectively, when water included rainfall, irrigation, and livestock drinking. Including lucerne reduced the water footprint of beef weight gain thanks to its ability to alleviate protein deficiency in the associated grasses, thus diminishing an important source of inefficiency in converting feed to meat.

Introduction

The Ogallala aquifer, one of the world's largest sources of freshwater used for irrigation, lies beneath semi-arid parts of eight states of the U.S. Great Plains from southern South Dakota southward to Texas. The high agricultural output and economic value of the irrigated Ogallala region is threatened by rapid groundwater declines (Hornbeck & Keskin, 2014; Steward et al., 2013). This includes the southern High Plains of northwest Texas and eastern New Mexico, where cotton, maize, sorghum, peanut, beef, and dairy are prominent agricultural commodities. The southern High Plains are particularly vulnerable to massive declines in economic productivity because the Ogallala aquifer is essentially nonrenewable thanks to minimal recharge and high evaporation rates. Quantifying the virtual water inherent in the production of a unit of commodity leads to deeper understanding of how to capture the highest value of the limited water resource (Williams & Al-Hmoud, 2015).

The water footprint of production from field to farm gate is a partial calculation in life cycle analyses of water use in commodity production. In analysis of water requirements of crops, a common measurement is the water use efficiency, or kg of commodity produced per cubic meter of water received from irrigation, rainfall, and in-season change in soil moisture, or water used in evapotranspiration. Water footprint can be expressed as the reciprocal of water use efficiency, therefore cubic meters of water received or evapotranspired per kg of commodity produced. The concept of water footprint of meat production is of particular interest in analyses of virtual water because meat, especially beef, has a very large water footprint relative to crops grown for human food (Mekonnen & Hoekstra, 2012), and its production involves large greenhouse gas (or carbon) footprints (White & Hall, 2017).

Water footprint estimates for beef and feed production

Water footprint estimates for beef production have been published for global, U.S., and for a portion of the Rolling Plains area of the southern Great Plains of the U.S., but the types of virtual water and types of animal product in those estimates varied, making comparisons difficult. The global analysis by Mekonnen and Hoekstra (2012) aggregated green (directly sourced from precipitation), blue (irrigation from stored sources), and grey (water needed to mitigate environmental degradation). They calculated water footprints of 10.2 m^3/kg of beef carcass weight when fattened on a high-concentrate diet (86% green water) vs. 21.8 m^3/kg on a pasture-only diet (96% green water). On a national scale, Beckett and Oltjen (1993) calculated the water footprint for boneless beef as 3.68 m^3/kg, but that only included blue water and water used in meat processing. In the southern Great Plains of the U.S., Kannon, Osei, Gallego, and Saleh (2017) considering only green water (evapotranspired water) in modeling the water footprint for grass production, but did not extend the calculation to beef output. Estimates varied from around 0.5 m^3/kg of feed dry matter for native range grasses to 1.0 m^3/kg for introduced (improved) grasses.

Most of the maize produced in the southern Great Plains goes to feeding cattle in confinement, largely for fattening beef (Auvermann & West, 2017). Water footprints for maize grain averaged 0.75 m^3/kg and for forage crops ranged from 2 to 5 m^3/kg (Williams & Al-Hmoud, 2015). The authors did not extend the water footprint estimate to include the conversion to beef. Rotz, Asem-Hiablie, Dillon, and Bonifacio (2015) reported a water footprint from birth to fattened product of 3.1 m^3/kg of carcass weight in the southern Great Plains, but only considered blue water. It is clear that substantial gaps in our knowledge of water footprint of beef production exist for a water-declining region, especially with respect to identifying management practices which enhance the efficiency of conversion of feed to marketable beef.

Effect of pasture legume inclusion on water footprint of stocker beef cattle

We provide an example here of a 3-year trial comparing two perennial-grass pasture systems grazed by beef stocker cattle for their water footprints, including green and blue water for forage production and directly for drinking. The two systems involved grasses growing without a legume but receiving a modest amount of nitrogen fertilizer and grasses growing with the legume lucerne (alfalfa) and receiving no nitrogen fertilizer. Details of trial set-up and animal productivity are reported by Baxter, West, Brown, & Green (2017a), and complete results of water footprint comparisons are reported by Baxter, West, Sarturi, Brown, and Green (2017b). Irrigation and drinking water was supplied from the Ogallala aquifer. Irrigation was limited on those paddocks containing Old World bluestem [*Bothriochloa bladhii* (Retz) S.T. Blake] and bluestem-lucerne mixtures to 230 mm annually, and to 300 mm annually on pastures containing predominantly lucerne. We summarize four methods of calculating water footprint to reflect the options of converting or not converting excess (ungrazed) forage into calculated animal liveweight gain, and calculating with and without precipitation (i.e., blue water only). Calculation of grey water was not done because there was no runoff from the pasture surface (0 to 1% slope), and nitrate leaching below the root zone would be negligible because of the restricted rooting depth.

The inclusion of lucerne reduced the water footprint of beef liveweight gain when both water sources (green + blue) were included in the calculation (Table 1), with values ranging from around 21 to 33 m^3/kg. Converting the ungrazed hay into predicted animal weight gain increased the denominator in the water footprint ratio, thereby lowering the value for the grass-alone treatment because there was much underutilized grass. Cattle more completely utilized the forage in the grass-legume system. Estimating the additional liveweight gain from the unutilized forage (hay) more completely accounts for potentially usable forage in the production system than if it were ignored. The presence of legume in the system lowered the water footprint relative to grass alone because of the greater rate of animal weight gain where lucerne was part of the system (0.94 vs. 0.79 kg/head/day, respectively) and greater weight gain per land area (188 vs. 118 kg/ha, respectively). The main attribute explaining the higher animal weight gain on the grass-legume treatment was the doubling of crude protein concentration of the available forage (14.4% vs. 7.0% for grass-alone). Therefore, the addition of lucerne removed a significant limiting factor to animal conversion of grazed forage into liveweight gain.

When water input from rainfall was not included in the calculations, water footprint values were an order of magnitude lower, ranging from 2.4 to 6.9 m^3/kg (Table 1), which more closely aligns these estimates with those of Beckett and Oltjen (1993) and Rotz et al. (2015), both of which omitted green water. This level of water footprint indicates that rainfall was by far the major water input, in agreement with Mekonnen and Hoekstra (2012). In our trial, adding the legume did significantly lower the water footprint in the non-converted method, but increased the value in the converted method. We trace that anomaly to the differential amounts of hay between the two

Table 1. Water footprint (m^3/kg) of two grazing systems with respect to non-converted and converted liveweight gain, with and without effective rainfall; data averaged across 3 years and 3 replicates. In all comparisons, grazing systems were significantly different ($P < 0.01$). Adapted from Baxter et al. (2017b).

		Grazing system treatment	
Water inputs	Conversion of hay to extra liveweight gain[a]	Grass-only	Grass-legume
Rainfall[b] + irrigation + drinking water	Non-converted	32.8	21.8
	Converted	28.2	25.4
Irrigation + drinking water	Non-converted	3.3	2.4
	Converted	3.7	6.9

[a] If non-converted, beef liveweight gain was calculated only as that amount observed in the grazing trials, not converting harvested hay into predicted (additional) liveweight gain. If converted, beef liveweight gain was increased by the amount calculated to occur if the harvested hay was fed to the same weight and class of cattle in the trial, based on nutrient contents of the hay. Methods used to convert hay to weight gain are available from the author.

[b] Rainfall was calculated as 'effective' rainfall, using a model which subtracts runoff according to soil water infiltration rate and storm intensity (Baxter et al., 2017b).

systems, which may have introduced error in the conversion to estimated liveweight gain.

In summary, including the legume lucerne in the pasture system lowered the water footprint in three out of four methods of calculation, and consistently did so when green and blue water inputs were included. The superior nutritional quality of the lucerne over the grass components resulted in greater animal growth, indicating that aspects of animal management for efficient use of all inputs must be optimized to cause maximum efficiency in conversion of water to beef animal growth. This was true in the case of adding lucerne to the system despite the fact that lucerne is notorious for its high requirement for water to reach high forage yields. Our deficit-irrigation management method held lucerne to low to moderate growth rates, and forced it to extract water from deep in the soil profile. We conclude that leveraging the exceptionally high quality of lucerne and its ability to persist and contribute enough to forage production under limited water input can contribute to lowering the water footprint of the stocker phase of beef production. This paper highlights the complexity of quantifying water footprint of beef production in pasture systems partly because of the high variations in forage utilizations and nutritional qualities of forages.

Acknowledgments

We appreciate the assistance of J. Sarturi, C. M. Brown, and P. E. Green in data collection and analysis. Lisa Baxter was partially supported by the Helen DeVitt Jones scholarship through the Texas Tech University College of Agricultural Sciences and Natural Resources. Field research was partially supported by U.S. Department of Agriculture SARE grant no. LS14-261. This paper was given at the workshop on Virtual Water in Agricultural Products: Quantification, Limitations and Trade Policy (Lincoln, Nebraska, USA, 14-16 September 2016), which was sponsored by the OECD Co-operative Research Programme: Biological Resource Management for Sustainable Agricultural Systems whose financial support made it possible for Charles West to participate in the workshop.

Disclosure statement

No potential conflict of interest was reported by the authors.

Funding

This work was supported by U.S. Department of Agriculture Southern SARE grant no. LS14-261.

References

Auvermann, B., & West, C. (2017). Ogallala region livestock systems. *Colorado Water, 34*(6), 33–35.

Baxter, L. L., West, C. P., Brown, C. P., & Green, P. G. (2017a). Stocker beef production on low-water input systems in response to legume inclusion. I. Forage and animal responses. *Crop Science, 57*, 2294–2302.

Baxter, L. L., West, C. P., Sarturi, J. O., Brown, C. P., & Green, P. G. (2017b). Stocker beef production on low-water input systems in response to legume inclusion. II. Water footprint. *Crop Science, 57*, 2303–2312.

Beckett, J. L., & Oltjen, J. W. (1993). Estimation of the water requirement for beef production in the United States. *Journal of Animal Science, 71*, 818–826.
Hornbeck, R., & Keskin, P. (2014). The historically evolving impact of the Ogallala aquifer: Agricultural adaptation to groundwater and drought. *American Economic Journal: Applied Economics, 6*(1), 190–219.
Kannon, N., Osei, E., Gallego, O., & Saleh, A. (2017). Estimation of green water footprint of animal feed for beef cattle production. *Water Resources and Industry, 17*, 11–18.
Mekonnen, M. M., & Hoekstra, A. Y. (2012). A global assessment of the water footprint of farm animal products. *Ecosystems, 15*, 401–415.
Rotz, C. A., Asem-Hiablie, S., Dillon, J., & Bonifacio, H. (2015). Cradle-to-farm gate environmental footprints of beef cattle production in Kansas, Oklahoma, and Texas. *Journal of Animal Science, 93*, 2509–2519.
Steward, D. R., Brussa, P. J., Yang, X., Staggenborg, S. A., Welch, S. M., & Apley, M. D. (2013). Tapping unsustainable groundwater stores for agricultural production in the High Plains Aquifer of Kansas, projections to 2110. *Proceedings of the National Academy of Sciences, 110* (37), E3477–E3486.
White, R. R., & Hall, M. B. (2017). Nutritional and greenhouse gas impacts of removing animals from US agriculture. *Proceedings of the National Academy of Sciences, 114*(48), E10301–E10308.
Williams, R. B., & Al-Hmoud, R. (2015). Virtual water on the Southern High Plains of Texas: The case of a nonrenewable blue water resource. *Natural Resources, 6*, 27–36.

Tradeoffs in the water-energy-food nexus in the urbanizing Asia-Pacific region

Makoto Taniguchi, Naoki Masuhara and Shun Teramoto

ABSTRACT

In order to evaluate the potential benefit and negative impacts of urbanization on the water-energy-food nexus, analyses have been made in Asia-Pacific countries with production-consumption relationships of water-energy-food nexus. Negative correlations have been found between urbanization rates and energy consumption rates for water and food production. This is because of the difference in decision making processes of resource development as urbanization progresses. Some positive relationships have been found between water consumption for energy production and Gross Domestic Product (GDP), and between water consumption for food production and hydro-meteorological conditions such as dry or wet countries.

Introduction

Water, energy, and food – the most fundamental resources for human beings – tie into a nexus through tradeoffs and connections in environment and society. Increases in population and changes in lifestyle that intensify consumption exert increasing demands on the nexus while climate changes affect the supply, making the water-energy-food nexus one of the most important global environmental and sustainability issues threatened by climate and societal changes (Bhaduri, Ringler, Dombrowsky, Mohtar, & Scheumann, 2015; Daher & Mohtar, 2015; Future Earth, 2014; Hoff, 2011; Mohtar, 2015; Mohtar & Lawford, 2016; Taniguchi, Masuhara, & Burnett, 2015; Van Vliet, Wiberg, Leduc, & Riahi, 2016).

Problems with the water-energy-food nexus are especially prevalent in areas where conflicts exist among demands on these three resources, as well as where the needs of resource suppliers develop when it comes to their resource use. This is true in particular for the densely populated urbanized regions with different types of demands for resources. The problem on the demand side of the nexus, such as an urbanized area with its larger population, has more potential for distribution conflicts due to consumption and waste generation. In contrast, the problem on the supply side is more limitation for the development of resource production because of dependency on resource availability.

The opinions expressed and arguments employed in this publication are the sole responsibility of the authors and do not necessarily reflect those of the OECD or of the governments of its Member countries.

Supplementary material for this article can be accessed at https://doi.org/10.1080/02508060.2018.1516104.

Therefore, the relationships between urban–rural as demand–supply, and water-energy-food nexus as consumption-production, overlap, making optimal decisions for resource use extremely difficult, but necessary for societies to become more sustainable.

The Asia-Pacific region also has one of the fastest rates of economic growth and resource consumption. This area contains 42% of the global population and is responsible for 58% of GDP and 63% of carbon emissions (IEA, 2012; Taniguchi et al., 2015; The World Bank, 2014). According to a previous study on water, energy, and food production in all 32 Asian-Pacific countries (Taniguchi et al., 2015), total water production (available water) in this area was 1.82×10^{12} m^3 in 2010 (FAO, 2014a) and 46.6% of global water production. Total energy consumption was 7.55×10^9 t of oil equivalent in 2010 (IEA, 2014) which is 61.2% of global energy consumption. Total food production was 2.37×10^9 t in 2010 (FAO, 2014a, 2014b), which is 54.6% of global food production. Therefore, this area covers about the half of all the world's production of water, energy, and food.

Water, energy, and food resources in the Asia-Pacific region are important on a global scale, yet these three resources are usually treated separately. Taniguchi et al. (2015) analyzed three resources in an integrated manner in terms of meeting security needs. They identified the sources and components of the three resources and evaluated their magnitude, diversity, and self-sufficiency in this region.

The objective of this study is to build on this work to evaluate the relationships between the water-energy-food nexus and urbanization, including the impacts of urbanization on the water-energy-food nexus. In particular, the relationship between production and consumption of resources, including water consumption for energy production, energy consumption for water production, food consumption for energy production, energy consumption for food production, and water consumption for food production are evaluated in the Asia-Pacific region.

Data and methods

Each country's consumption and production of water, energy, and food in 2010 were collected from the FAO (2014a, 2014b) and IEA (2014), and demographic data such as population, GDP at market prices (in current US$), World Development Indicators of the World Bank, UN data, and urbanization rate were collected from the United Nations (2016). Definitions of urbanization are generally those used by national statistical offices in carrying out the latest available census, according to the explanation of the Population Division, Department of Economic and Social Affairs, United Nations. Data from 11 different countries surrounding the Asia-Pacific region in 2010 were examined: New Zealand, Australia, Indonesia, the Philippines, China, Japan, Russia, Canada, the United States, Mexico, and Peru.

In this study, 'water for energy' means the amount of water consumption by evaporation from hydropower plants and cooling water use for nuclear and thermal power plants (unit requirement × power generation amount). 'energy for water' means the energy consumption for water transportation, distribution and pumping. 'food for energy' means fuel ethanol consumption (liter)/unit requirement. 'energy for food' means energy unit requirement × food production. 'water for food' means water unit requirement × food production (cereals, vegetables, fruits, livestock and cultured fish). The details of the data sources are found in Supplement Information-1.

Data on energy consumption for water pumping in China was not found. Therefore, the total energy consumption for water production in China was underestimated. In the case of water data from FAO (2014a), the water volume shows the available amount of water in each country, so if water is reused from upstream to downstream, it may be double or triple counted. There are also no data on water consumption for food production in New Zealand.

There are other limitations of the data available for the water-energy-food nexus. For instance, food waste that pollutes water, and energy is that consumed cleaning water contaminated by agricultural fertilizers and pesticides in agriculture were not included in this study.

Results

The results are categorized into three tradeoffs for the following discussion on the relationship between the water-energy-food nexus and urbanization: (1) water and energy; (2) energy and food; and (3) water and food. The numbers are summarized in Table 1 of supplementary file.

Tradeoffs between water and energy

Tradeoffs between water and energy occur between water consumption for energy production and energy consumption for water production. Water consumption for energy production includes use in hydro-power plant generation and in cooling systems for thermal and nuclear power plant generations. Energy consumption for water production occurs through water allocations including water pumping and water allocations. Both water and energy consumption depend on the efficiency of the infrastructure in each country.

Water consumption for energy production in 11 Asia-Pacific countries is shown as bar graphs in Figure 1 for hydropower generation and cooling for both nuclear and thermal power plants. The proportions of the three different types of water consumption for energy production to the total water consumption for energy production are also shown as pie charts in Figure 1. The 2010 total water consumption for energy production is shown for each country in bar graphs at the bottom of Figure 1. The largest total volume of water consumption for the energy productions was found in the United States of America (821 x 10^9 $m^{3/}y$), followed by China (749 x 10^9 $m^{3/}y$). The largest volume of water consumption for hydropower plants is found in China (7% of total water use), although the highest proportion of water consumption for the hydropower plants was in New Zealand (41%), followed by Canada (33%), and Peru (32%). The largest volume of water consumption for the cooling system of nuclear power generation is found in the United States of America (21% of total water use), although the highest proportion of water consumption for nuclear power plants cooling was in Japan (29%), followed by Canada (26%), and the United States of America (21%). However, this situation in Japan has changed since the Great East Japan Earthquake on 11 March 2011, because the operations of all nuclear power plants in that country were

Figure 1. Water consumption for energy production. The amounts of water used for hydropower, cooling for nuclear power plants and thermal power plants, and the proportions of three nexus water consumptions to the total water consumption.

stopped. As for water consumption for the cooling system of a thermal power plant using oil, coal, or gas power generation, the United States of America consumed the largest volume of water for cooling, followed by China. The highest proportion of water consumption for the cooling system of a thermal power plant was in Australia (98%), followed by Indonesia (96%), the Philippines (95%), and Mexico (92%).

Energy consumption for water production, including water allocations as water pumping and water transports, is shown in Figure 2. In addition, the proportions for the two different types of energy consumption for water production are depicted in pie charts. The 2010 totals for energy consumption for water production in each country are shown in bar graphs at the bottom of Figure 2. The United States used the most energy for water allocation for both water pumping and water transport, followed by China. In descending order, Russia, Canada, Japan, and Indonesia had the largest energy consumption for water allocation. Energy consumption for water pumping alone is also largest in the US, followed by Japan. The highest rate of energy consumption for water pumping was in Japan (7%), followed by the United States (4%). On the other hand, the highest rate of energy consumption for water transport was in the Philippines. The largest total amount of energy consumption for water production was found in United States, followed by China, Russia, Canada, and Japan.

Figure 2. Energy consumption for water production. The amount of energy consumption for water pumping and transport, and the proportions of two types of nexus energy consumption to the total energy consumption.

Tradeoffs in the energy and food nexus

Tradeoffs between energy and food occur as food consumption for energy production and energy consumption for food production. Food consumption for energy production consists of biomass energy. This study uses the data of potential biomass energy from corn, wheat, rice, and root sugars in each country (Taniguchi et al., 2015). In contrast, energy consumption for food production occurs though growing cereals, vegetables, fruits, meats, and fisheries, as well as through all food transports. Energy consumption for growing crops includes only inputs to production process, such as tractors, harvesting and fertilizers. Energy consumption for food trade is not included in this analysis.

The weights of food consumption for biomass energy production are shown in Figure 3. The largest food consumption for energy production was in the United States, followed by China, Canada, and Australia. There was no 2010 biomass production in Russia and Indonesia.

Energy consumption amounts for foods such as cereals, vegetables, fruits, livestock (meats), and fisheries are shown in bar graphs of Figure 4. The proportions of five different types of energy consumption for food production are also shown as pie charts in Figure 4. The 2010 total energy consumption for food production in each country is shown in a bar graph at the bottom of Figure 4. The largest energy consumption for all

Figure 3. Food consumption for the energy production. Potential biomass from corn, wheat, rice, and root sugars, and the proportions of food consumption for the energy production to the total food consumption.

Figure 4. Energy consumption for food production.

food production was in China, followed by the United States, but the second largest energy consumption for fisheries was in Indonesia. For energy consumption for vegetables and fruits, the highest totals were in the United States, followed by Mexico and the Philippines.

The largest proportion of energy consumption for cereal production was in Canada (74%), followed by the United States (72%) and Australia (72%). For vegetables, the largest proportion of energy consumption was in China (9%), followed by Japan (6%). In contrast, the largest proportion of energy consumption for fruit production was in Peru (16%), followed by the Philippines (13%) and Mexico (10%). For livestock (meats), the largest proportion of energy consumption was in New Zealand (66%), followed by Japan (37%) and Peru (36%). In contrast, the largest proportion of energy consumption for fishery production was in the Philippines (24%), followed by Indonesia (23%) and China (22%).

Regarding the proportion of energy consumption for each food product, the countries in the Western Pacific Ocean such as the Philippines, Indonesia, China and Japan consumed more energy for fisheries. In North America, however, the US and Canada consumed more energy for cereal production. Some tropical countries such as the Philippines and Mexico have a larger proportion of energy consumption for fruit production compared to other countries.

Tradeoffs in the water and food nexus

Tradeoffs between water and food occur as water consumption for food production. Water is not produced directly from food, therefore no food consumption for water production in the water and food nexus. Water consumption for food production, such as for cereals, vegetables, fruit, and livestock (meats) and fisheries, is shown for each country in bar graphs of Figure 5. The proportions of five different types of water consumption for food production are also shown in pie charts in Figure 5. The 2010 total water consumption for the food production in each country are shown in bar graphs at the bottom of Figure 5. The largest water consumption for cereals was in China, followed by United States, Indonesia, and Russia. For vegetable production, the largest water consumption was in China, followed by the United States and Indonesia. The largest water consumption for fruit was also in China as well as tropical countries such as the Philippines, Indonesia, and Mexico, which is the same as energy consumption for fruit production. Therefore, those four countries are both water and energy intensive for fruit production. The largest water consumption for meat production was in the United States and China, followed by Mexico and Russia. For fishery production, the largest water consumption is found in China, followed by Indonesia, which is similar to the energy consumption for fishery production. There was no fishery data for New Zealand.

The highest proportions of water consumption for the cereal productions are found in Canada (63%), Indonesia (59%), the Philippines (59%), Russia (55%), the United States (49%), and Australia (46%) (Figure 5). On the other hand, the proportion of water consumption for vegetable production was less than 3% for all countries. The highest proportions of water consumptions for fruit production was found in the Philippines (16%), Peru (11%), Mexico (8%), and Indonesia (6%). For meat production, the highest proportion of water consumption was in Mexico (51%) and the United

Figure 5. Water consumption for food production.

States (46%). In contrast, the highest proportion of water consumption for fishery production was in China (77%), Peru (36%), Japan (33%), and Indonesia (27%).

Five tradeoffs – water for energy, energy for water, food for energy, energy for food, and water for food – in 11 Asia-Pacific countries are summarized (Table 1, supplementary file), as well as population, GDP, and urbanization rates. Urbanization rate is defined as the size of population in an urban area divided by total population. Urbanization and industrialization in particular in developing countries, increase demand for resources, including water, energy, and food. Therefore, urbanization is one of driving forces for a changing the water-energy-food nexus (Satterthwaite, McGranahan, & Tacoli, 2010; Taniguchi et al., 2015).

The relationships between urbanization and nexus rates of 11 countries are shown in Figure 6 as (1) the percentage of water consumption for the energy production to the total water consumption, (2) the percentage of energy consumption for water production to the total energy consumption, (3) the percentage of food consumption for the energy production to the total food consumption, (4) the percentage of energy consumption for food production to the total energy consumption, and (5) the percentage of water consumption for the food production to the total water consumption.

Discussion

Slightly negative relationships were found between the urbanization rate (x) and the proportion (Y) of energy for water (Figure 6(b)) and between the urbanization rate (x)

Figure 6. Relationships between urbanization and nexus rates of (a) water for energy, (b) energy for water, (c) food for energy, (d) energy for food, and (e) water for food.

and the proportion (Y) of energy for food (Figure 6(d)). China does not include any data for energy consumption for water pumping, therefore the plot of China in Figure 6 (b) may be located in a higher proportion of energy for water with the same urbanization rate. In contrast, the relationships between urbanization rates and water for energy (Figure 6(a)) and water for food (Figure 6(e)) are not clear. Note the water-for-food nexus rate in China exceeds 100% because the water is reused for food production within the basin. One potential reason for the difference between the rates of energy (for water and food) and of water (for energy and food) may be the difference in the decision making process for development of the resources. In the case of energy developments, there is less conflict and tradeoff for use as nexus during urbanization because the energy production facilities such as thermal power plant, hydropower plant and nuclear power plants which use point sources of energy, are made after decision making regarding the use of energy. In contrast, river water and groundwater are spatial resources (Taniguchi & Hiyama, 2014), and the water resources existed before making decision regarding the use for energy or food as nexus. Therefore, there are many alternative uses, and potential for conflicts between water for energy or food due to increased water demand from urbanization.

However, the lack of land for energy development post-urbanization causes longer distance between energy source and power plant, thus consuming more energy for transportation. For instance, many thermal power plants are located near the coastal zone where fuel transportation is less costly. Another conflict on the location of the power plants is that many power plants have to be placed near water sources for cooling. For instance, there is less conflict on water use for cooling or others in island countries because seawater near the coast is used for cooling. On the other hand, there is more conflict on water use for cooling or others in inland countries because fresh water is the only available source for cooling. Therefore, there are still many factors on relationships between urbanization and nexus rates that require further studies.

Theoretically, potential regression lines/curves should cross the origin of the graph because there is no nexus at the beginning of the urbanization. Therefore, the negative correlations of energy for water (Figure 6(b)) and food (Figure 6(d)) may show the part of potential regression curves. If so, during early urbanization stages (lower urbanization rates), urbanization has a negative impact on energy consumption for water or food, which means an increase in energy consumption for water or food. However, urbanization may have a benefit for energy consumption for water or food during later urbanization stages (higher urbanization rates, Figure 6(b,d)). This is still a preliminary argument with limited data and additional studies are needed.

According to the summary of water-energy-food nexus in 11 Asia-Pacific countries (Table 1, supplemental file), the highest rates of water for energy to total water consumption are found in United States, China, and Japan. This happens in the countries with higher GDP numbers. Regarding the food consumption for energy production, only the United States has a higher rate (more than quarter) to the total food production, which is a result of United States energy policy. In contrast, higher rates of water consumption for food production to the total water consumption are found in Australia, China, and Mexico, which are relatively dry countries.

Conclusion

The water-energy-food nexus in 11 Asia-Pacific countries, including developed and developing countries, is analyzed in terms of urbanization and production-consumption relationships. Negative correlations were found between urbanization rates and proportions of energy consumption for water and food production. In contrast, no relationships were found between urbanization rates and rates of water consumption for energy and food production. This may be explained by the difference in decision making processes for resource development during urbanization. Energy developments occur after making decisions of the purpose of the energy use. In contrast, water resources exist as spatial resources before making decisions for the water use. Therefore, alternative water use may likely occur due to increases in demands of water during urbanization. The negative relationship between urbanization rates and rates of energy consumption for water and food production means that there are benefits of urbanization for the nexus. The relationship between urbanization rate and the proportion of food consumption for energy production is also not clear. Some nexus relationships of water are found between water consumption for energy production and GDP as well as for water consumption for food production and for hydro-meteorological conditions, such as in contrasting dry and wet countries. This paper contributes to the understanding of how water-energy-food production-consumptions are related, and the potential benefits and negative impacts of urbanization on the water-energy-food nexus.

Acknowledgments

The authors would like to acknowledge the Research Institute for Humanity and Nature (RIHN), Japan, for the financial support for this study. We would like also to acknowledge Dr. Joshua Newell, School of Natural Resources and Environment, University of Michigan, for his useful and insightful comments. This paper was given at the workshop on Virtual Water in Agricultural Products: Quantification, Limitations and Trade Policy (Lincoln, Nebraska, USA, 14-16 September 2016), which was sponsored by the OECD Co-operative Research Programme: Biological Resource Management for Sustainable Agricultural Systems whose financial support made it possible for the author to participate in the workshop.

Disclosure statement

No potential conflict of interest was reported by the authors.

Funding

This work was supported by the Research Institute for Humanity and Nature [14200097].

References

Bhaduri, A., Ringler, C., Dombrowsky, I., Mohtar, R., & Scheumann, W. (eds.). (2015). Sustainability in the water-energy-food nexus (Special Issue). *Water International*, *40*(5–6), 723–732.

Daher, B. T., & Mohtar, R. H. (2015). Water–Energy–Food (WEF) Nexus tool 2.0: Guiding integrative resource planning and decision-making. *Water International*, *40*, 748–771.

FAO. (2014a). *FASTAT*. Rome: Italy Food and Agriculture Organization of the United Nations. Retrieved from http://faostat3.fao.org/home/index.html

FAO. (2014b). *FishStatJ*. Rome: Italy Food and Agriculture Organization of the United Nations. Retrieved from www.fao.org/fishery/statistics/software/fishstatj/en

Future Earth ICSU. (2014). *Future earth strategic research agenda 2014*. Paris: France International Council for Science Union. Retrieved from www.futureearth.org/sites/default/files/strategic_research_agenda_2014.pdf

Hoff, H. (2011). *Understanding the nexus: Background paper for the Bonn 2011 conference: The water, energy, and food security nexus*. Stockholm, Sweden: Stockholm Environment Institute. Retrieved from http://wef-conference.gwsp.org/fileadmin/documents_news/understanding_the_nexus.pdf

IEA. (2012). *CO2 emissions from fuel combustion (2012 Edition)*. Paris: France International Energy Agency.

IEA. (2014). *Energy balances of OECD countries and non-OECD countries*. Paris: France International Energy Agency. Retrieved from www.iea.org/statistics/topics/energybalances

Mohtar, R. H. (2015). Ven Te Chow memorial lecture: Localizing water and food security. *Water International*, *40*, 559–567.

Mohtar, R. H., & Lawford, R. (2016). Present and future of the water-energy-food nexus and the role of the community of practice. *Journal Environment Studies & Sciences*, *6*(1), 192–199.

Satterthwaite, D., McGranahan, G., & Tacoli, C. (2010). Urbanization and its implications for food and farming. *Philosophical Transactions of the Royal Society B: Biological Sciences*, *365* (1554), 2809–2820.

Taniguchi, M., & Hiyama, T. (2014). *Groundwater as key of adaptation to the climate change*. Tokyo: Japan Springer.

Taniguchi, M., Masuhara, N., & Burnett, K. (2015). Water, energy, and food security in the Asia Pacific region. *Journal of Hydrology: Regional Studies*. doi:10.1016/j.ejrh.2015.11.005

United Nations. (2016). *UN data*. New York: US. Retrieved from http://data.un.org/

Van Vliet, M. T. H., Wiberg, D., Leduc, S., & Riahi, K. (2016). Power-generation system vulnerability and adaptation to changes in climate and water resources. *Nature Climate Change*, *6*, 375–381.

The World Bank. (2014). *World development indicators*. Washington, DC: US. Retrieved from http://data.worldbank.org/data-catalog/world-development-indicators

Index

Note: **Bold** page numbers refer to tables; *italic* page numbers refer to figures.

For Product Safety Concerns and Information please contact our EU
representative GPSR@taylorandfrancis.com
Taylor & Francis Verlag GmbH, Kaufingerstraße 24, 80331 München, Germany

www.ingramcontent.com/pod-product-compliance
Ingram Content Group UK Ltd.
Pitfield, Milton Keynes, MK11 3LW, UK
UKHW052212190926
475702UK00006B/978

9780367408077